WELDER'S HANDBOOK

A GUIDE TO PLASMA CUTTING, OXYACETYLENE,
ARC, MIG AND TIG WELDING

RICHARD FINCH, S.A.E., A.W.S.

HPBOOKS

HPBooks
Published by the Penguin Group
Penguin Group (USA) Inc.
375 Hudson Street, New York, New York 10014, USA

Penguin Group (Canada), 90 Eglinton Avenue East, Suite 700, Toronto, Ontario M4P 2Y3, Canada
(a division of Pearson Penguin Canada Inc.)
Penguin Books Ltd., 80 Strand, London WC2R 0RL, England
Penguin Group Ireland, 25 St. Stephen's Green, Dublin 2, Ireland (a division of Penguin Books Ltd.)
Penguin Group (Australia), 250 Camberwell Road, Camberwell, Victoria 3124, Australia
(a division of Pearson Australia Group Pty. Ltd.)
Penguin Books India Pvt. Ltd., 11 Community Centre, Panchsheel Park, New Delhi—110 017, India
Penguin Group (NZ), 67 Apollo Drive, Mairangi Bay, Auckland 1311, New Zealand
(a division of Pearson New Zealand Ltd.)
Penguin Books (South Africa) (Pty.) Ltd., 24 Sturdee Avenue, Rosebank, Johannesburg 2196, South Africa

Penguin Books Ltd., Registered Offices: 80 Strand, London WC2R 0RL, England

While the author has made every effort to provide accurate telephone numbers and Internet addresses at the time of publication, neither the publisher nor the author assumes any responsibility for errors, or for changes that occur after publication. Further, publisher does not have any control over and does not assume any responsibility for author or third-party websites or their content.

Welder's Handbook

First edition: April 2007

ISBN: 978-1-55788-513-5

PRINTED IN THE UNITED STATES OF AMERICA

10 9 8 7 6 5 4 3

NOTICE: The information in this book is true and complete to the best of our knowledge. All recommendations on parts and procedures are made without any guarantees on the part of the author or the publisher. Tampering with, altering, modifying, or removing any emissions-control device is a violation of federal law. Author and publisher disclaim all liability incurred in connection with the use of this information.

CONTENTS

ACKNOWLEDGMENTS

First and Second Edition Acknowledgments

Mr. Rick Wuchner of So-Cal Air Gas, Ventura, CA; Mr. Jeff Noland of HTP America, Inc., Arlington Heights, IL; Mr. Bill Berman of Daytona Mig, Inc., Daytona Beach, FL; Mr. Ed Morgan of Lincoln Electric Co., Santa Fe Springs, CA; Mr. Ray Snowden of Allan Hancock College, Santa Maria, CA; Mr. Seth Hammond and Mrs. Tannis Hammond of Specialty Welding, Goleta, CA; Mr. Tom Giffen and Mr. Ron Chase of Western Welding, Goleta, CA; Mr. Dave Williams of Williams Lo-Buck Tools, Norco, CA; Mr. Michael Reitman of United States Welding Corporation, Carson City, NV; Ms. Darlene Tardiff of Harbor Freight Tools, Inc., Camarillo, CA; Mr. Phil Pilsen of Pro-Tools, Tampa, FL; Mr. Henry Hauptfuhrer of The Eastwood Co., Malvern, PA; Ms. Lisa West of Smith Equipment, Watertown, SD; Mr. Dale Wilch of Dale Wilch Sales, Kansas City, MO; Mr. Hal Olcutt of Victor-Thermadyne, Denton, TX; Mittler Bros Tools, Huntsman Welding Helmets, Salt Lake City, UT; Mr. Dick Casperson of Miller Electric Co., Appleton, WI; and two friends who helped, Mr. Dale Johnson and Mr. Jerry Jones, both of Goleta, CA.

Third Edition Acknowledgments

In addition to the people that were named in the first and second editions of *Welder's Handbook*, I want to recognize other people too. These are: Scott Skrjanc, Events Marketing Manager, Lincoln Electric Co.; Jenny Ogborn, Photographic Dept, Lincoln Electric Co.; Mike Pankoratz, Miller Electric Co.; Ray Kudlak and Phil Baldwin of Cronatron Welding Supplies; Jason Woods, Alamogordo, NM; Jim Holder, Tularosa, NM; John Gilsdorf, Alamogordo, NM; Mike Reid, owner of Crown Welding and Industrial Supply of Alamogordo, NM; and Dottie Hammack, who runs Valley Welding and Gasses of Alamogordo, N.M.

THE WORLD DEPENDS ON WELDING

Welding is an art, and it is also a means to achieve an end result. For instance, welding is a process used to build automobiles and airplanes, and it is also a process used to build stronger buildings, highway bridges, hydroelectric power plants, and probably the chair you are sitting in while you read this page.

If you possess a television set and a refrigerator, spot welding was used to make those appliances. If you are reading this page in a bookstore or a library, the bookshelves probably utilize welded brackets and bases to hold the shelves in place. If every weld in the world should give way at once, our world would fall apart. Automobiles would fall into piles of metal in the parking lots, airplanes would fall from the sky in pieces, ships would come apart and sink, and even the bed you sleep in would fall to the floor.

But welding is a satisfying talent that almost anyone can develop. I really enjoy building things out of metal. The welding table that I designed and built for a project in this book is one of the most useful welding projects that I have ever completed. Regardless of your reason for wanting to learn to weld, or to be a better welder, you will surely want to build this table before you try any other projects because this welding table will support many of your future projects. If you plan to build lots of welded projects, then you may want to build a much larger welding table, up to 4' x 8' in size.

Learning to weld is similar to learning to play a piano. You can be playing "Chopsticks" with just a few minutes' instruction, but it takes a lot of practice to play Mozart on the piano. Likewise, you can learn to run a weld bead in just a few minutes, but you will have to spend time practicing before you build a real race car, airplane, or an ocean-going aluminum yacht. The projects that I have developed for you in Chapter 16 provide a low-pressure way to practice your welding before you tackle the really important projects.

This is a typical back driveway trailer building project, using a new Lincoln Invertec 205-T AC/DC welder that can do both TIG (Heli-Arc) and stick/arc welding. The project is also supported by a small Harris Gas welding and cutting torch and portable carrier. The trailer is upside down for fitting the wheels and axles.

A WORD ABOUT THE THIRD EDITION

This third edition of *Welder's Handbook* comes at an exciting time in the history of welding. Electronic technology has experienced a major change in many areas of our lives, in things like cell phones that fit in your shirt pocket, that can send and receive calls anywhere in the world, and in laptop computers that have more power than the school bus-sized computers that existed when the first edition of this book was written.

And the same is true for arc welding equipment that was the size of a refrigerator when the first edition was written, that now is about the size of a lunch pail and almost as portable. New arc welding machines now can also be plugged into 220 volt and 230 volt shop current and then simply be taken away in the trunk or front seat of your car and plugged in to 110 volt house current and used to weld just as well as they were doing in the shop environment. And all this new welding technology has occurred in the past ten years since the second edition of this book was written.

Other new technology has emerged in the form of television programs on the Learning Channel, on The Discovery Channel, and do-it-yourself shows, such as *Monster Garage* featuring Jessie James; *Overhaulin'*, featuring Chip Foose (I worked side-by-side with Chip for more than two years at the ASHA Corp. in Goleta, CA, in the mid 1990s); *American Chopper: Orange County Choppers*, featuring the Tuttle family; how-to shows featuring Boyd Coddington and other good shows that demonstrate how experts weld and form metal.

One important suggestion: If you don't know much about welding, then read the glossary in the back of this book for explanations of terms normally used in the welding trade, and refer to it anytime you don't understand a word in the book.

—*Richard Finch*

Chapter 1

METAL BASICS AND HEAT CONTROL

Tony Stewart's Nextel Cup race car gets some MIG (wire feed) welding repairs prior to the next race. Notice the cleanliness of the welding shop. Your welding shop should be clean too! Photo Courtesy Lincoln Electric Co.

There are many factors involved in producing good welds in metal: the equipment used, the filler metal used, the preparation of the parts to be welded, and especially, the correct application of heat to the weld. In manual (not automatic) welding, the welder is the artist who knows how much heat to apply to produce good welds. Before you can weld, you must learn how to control heat, and have some fundamental knowledge of the basic types of metals and their properties.

SOLID, LIQUID, GAS

My high school physics teacher, Larry Grundy, taught me that almost all matter on this earth is in three basic forms: (1) Solid (frozen), (2) Liquid (molten), and (3) Gas (vapor). At the time, I don't think that I believed him, but as I got older and tried things my own way, I began to see that he was right. I learned by experience that temperature, and especially heat control, plays a very important part in how the earth exists. And it really plays a very important part in welding. Note: When I was in high school, the 4th state of matter had not yet been explored to any extent, but the 4th state of matter is a serious thing today. It is called "plasma" and Webster's Dictionary defines it as "a collection of charged particles containing about equal numbers of positive and negative ions and electrons, exhibiting some of the same properties of a gas, but differing from a gas in being a good conductor." This is important when cutting metal with a plasma cutter as we do in welding today. Please refer to the glossary on page 149 for more information.

Example: Metal can exist in these three forms, just like ice. Steel is solid at or below its 2,700°F melting point. Ice is solid at or below its 32°F melting point. Heat the ice on a kitchen stove to 32°F and it becomes liquid water, then heat it to 212°F and it begins to boil and vaporize.

Heat steel to 2,786°F and it becomes liquid or molten, then continue to heat it to over 5,500°F and it begins to vaporize. In welding, when the liquid/molten *weld puddle* (see glossary), begins to solidify, we say it is freezing. Certain kinds of arc welding rods are called "fast freeze rods."

While you are learning to weld, keep the example of ice cubes on the kitchen stove in mind. If you get the metal too hot, it will vaporize, just like the ice cubes do. Study the charts in this chapter to get a better idea of the melting and boiling points (temperatures) of the many kinds of metals that are weldable.

COLOR CHANGES OF METALS

As you practice your welding, brazing and soldering skills, you will recognize the color changes in metals as the metal is heated and cooled. Study the charts in this chapter to better understand the temperatures that result in specific colors when you are heating steel.

Aluminum does not exhibit the same kind of color changes that steel does, but it goes from shiny to dull as it is heated, then to shiny again as it begins to melt.

Stainless steel does not go through the number of color changes that mild steel and carbon steel go through in increasing temperatures, but stainless steel does turn red just

TEMPERATURES OF SOLDERING, BRAZING & WELDING PROCESSES

Soldering, Lead Solder: 250–800°F (121–427°C)
Brazing, Brass and Bronze: 800–1,200°F (427–649°C)
TIG Welding: 5,000°F (2,760°C), Arc-Temp Variable
Oxyacetylene: 6,300°F (3,482° C) Flame Temp, Adjustable
Oxyacetylene Cutting: 6,300°F (3,482° C), Flame Temp, Adjustable
Arc Welding: 6000–10,000°F (3,316–5,538° C), Arc Temp
Plasma-Arc Cutting: 50,000°F (27,760° C), Arc Temp

Keep these working temperatures in mind. Each process—welding, brazing or soldering—is different. If you overheat the weld bead, you could "vaporize" your project! Master temperature control, and you will become a much better welder.

The welder in this picture is making a MIG (Wire Feed) welding repair to the front frame of this Nextel Cup Monte Carlo race car that is driven by Bobby Labonte. Photo Courtesy Lincoln Electric Co.

before it melts.

Similarly, brass and copper do not show the same color changes that steel does. Brass just gets lighter in color, then shiny as it melts; copper just gets red, then dark as it reaches melting temperature.

CONTROLLING HEAT

First, study the chart labeled *Temperatures of Soldering, Brazing and Welding Processes.* You will see a temperature range for each process. Soldering is the lowest temperature range, 250°F to 800°F. Brazing is the next highest temperature range at 800°F to 1,600°F, and Fusion welding at the melting point of the metal to be welded.

Next, study the chart *Weights, Melting Points & Boiling Points of Metals.* From this chart, you will begin to understand why you must control the heat in order to produce the results you want, which is good, strong metal joining techniques.

In every welding, brazing and soldering process that I describe and instruct in this book, I tell you to control the heat. Too cold will produce a weak bond or no bond at all, and too much heat will boil or vaporize the bond and even ruin your metal.

TYPES OF METAL

As you read this section, I strongly suggest you refer to the glossary for further definitions of terms you do not fully understand.

Before you can begin welding, you must know what kind of metal you are going to weld. Usually, but not always, you should weld the same kind of metals together. There will be cases where you will

want to weld stainless steel to low alloy carbon steel, or solder brass to aluminum. In the applicable chapters, I will show you how to do it. But first, study the following paragraphs and photos to find out how to identify metals.

Ferrous & Non-Ferrous Metals

Ferrous indicates that a metal has iron content, and that it is attracted to a magnet. Non-ferrous means that the metal doesn't have any iron content and that it is not magnetic. Cast iron, mild steel and chrome molybdenum steel are ferrous metals. Aluminum, brass, copper, gold, silver, lead and magnesium are non-ferrous metals and they are not magnetic and are not attracted to a magnet. In welding terms, ferrous and non-ferrous metals are considered to be "dissimilar metals" but they can be joined by various welding, brazing or soldering processes.

For instance, I'll show you a special Cronatron (www.cronatron.com) brand of solder that will join any and all metals if a tensile strength of the bond at 7,000 psi would be acceptable. In this book, I will also show you how to braze copper to cast iron and even how to do repair welds without heat. More on that later.

The non-heat welding process is called JB Weld and you will have to read the appropriate chapter of the book to find out how to do it. Now, let's become more familiar with the various types of metals.

WEIGHTS, MELTING POINTS & BOILING POINTS OF METALS

Metal	Weight lbs/ft^3	Melting Point °F (C)	Boiling Point °F (C)
Aluminum	166	1,217 (658)	4442 (2450)
Bronze	548	1,566–1832	(850–1000)
Brass	527	1,652–1724	(900–940)
Carbon	219	6,512 (3600)	
Chromium	431	3,034 (1,615)	
Copper	555	1,981 (1,083)	4,703 (2595)
Gold	1205	1,946 (1,063)	5,380 (2971)
Iron	490	2,786 (1,530)	5,430 (2999)
Lead	708	621 (327)	3,137 (1725)
Magnesium	109	1,100 (593)	
Manganese	463	2,300 (1260)	
Mild Steel	490	2,462–2786 (1,350–1,530)	5,450 (3049)
Nickel	555	2,645 (1,452)	4,950 (2732)
Silver	655	1,761 (960)	4,010 (2210)
Tin	455	449 (231)	4,120 (2271)
Titanium	218	3,263 (1795)	
Tungsten	1186	5,432 (3,000)	10,706 (5930)
Zinc	443	786 (419)	1,663 (906)
4130 Steel	495	2,550	5,500 (3051)

Melting points of metals vary widely. Use this chart to determine how hot you will need to get the metal in order to fusion weld it. If available, I have also included boiling temperatures for each metal, where they are available, a temperature you should surely avoid.

This illustrated thermometer will help you visualize the melting points of various metals. It will also give you an idea of how much heat you will need to weld various metals.

TITANIUM → 3250F
MILD STEEL → 2750F
COPPER →
GOLD →
BRASS →
ALUMINUM →
MAGNESIUM →
LEAD →
TIN →
HUMAN BODY →

3500F
3250F
3000F
2750F
2500F
2250F
2000F
1750F
1500F
1250F
1000F
750F
500F
250F
0F

Cast Iron—This metal is usually rough-textured because of how it is manufactured. A cast-iron part is formed by pouring molten iron into a sand mold, thus giving it the form and texture of the interior of the mold. Typical cast-iron parts include automotive engine blocks, exhaust manifolds, manual-transmission cases, older lawn mower and garden-tractor engines, and early-style farm equipment.

Where it is cut with a lathe, saw, grinder or whatever, cast iron usually has a gray, grainy appearance. When ground with a high-speed grinding wheel, red sparks are generated.

Cast iron is ferrous, or magnetic. It can be arc-welded with stick-type electrodes or brazed or fusion-welded with an oxyacetylene torch.

Forged Steel—This is a rough metal, but smoother than cast iron. It's used for most engine connecting rods, some crankshafts, axle shafts and some chassis components. Forging steel is done by hammering a red hot steel billet (see glossary) into the desired shape in a forging press. Machined, cut or ground, forged steel is light gray or silver inside. Grinding forged steel creates yellow or white sparks.

Forged steel is a ferrous metal. It can be welded with gas, arc, TIG, MIG methods. But, because steel forgings are intended for high-load/high-fatigue applications, such parts should be welded using the best methods available—TIG, or DC arc—if welded at all. Usually, damaged forged-steel parts should be replaced rather than welded.

Stainless Steel—Stainless is very smooth and hard. It's usually found in sheets, but can be cast. As the name implies, it will not stain or corrode easily. Stainless steel is used to make kitchen cutlery, pots and pans, exhaust systems for airplanes or autos. When ground with a high-speed abrasive stone, it does not give off sparks. Instead, stainless steel turns

Chrome moly steel is factory-marked for ease of identification. Here you see condition N (normalized), 4130-HR (4130 hot rolled), .095 x 29 (.095" thick x 29" wide) x 72 (72" long), OHT 144610 (heat number 144610 certification).

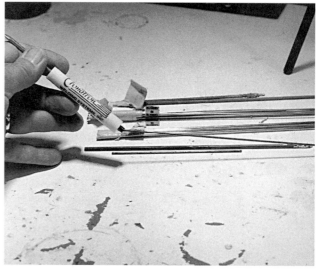

Small magnetic screwdriver is a handy way to test for ferrous or non-ferrous metals. In this picture, the magnet will not pick up brass welding rod because brass is non-ferrous, thus non-magnetic.

black where ground.

Stainless steel is an exception to the ferrous rule. It is steel, but it is usually non-magnetic. However, a few of the hundred or so different stainless steel alloys are magnetic. Note that some stainless steels do contain some amounts of ferrous metals such as iron and steel. Stainless steel welds best with TIG, but arc and gas welding can be used as well. It can also be gas-brazed.

Mild Steel—This is the most common of all metals, used in automobile bodies and chassis parts, lawn-mower handles, bicycle fenders, house furniture, filing cabinets—the list is endless. Cut or ground, it looks bright gray to silver. Grinding with a stone wheel generates a shower of yellow sparks.

Mild steel is ferrous. It can be welded by every method described in this book: gas, arc, TIG and MIG, and spot welding. It is very easy to work with.

Chrome Moly Steel—Chrome moly is also called low carbon steel because it contains some carbon for greater strength than mild steel but less than 0.04%. It is ferrous and can be heat treated after welding to 190,000 psi tensile strength. It welds best with TIG or oxyacetylene and should not be brazed because it has grains which can open to molten brass and often cracks as a result of brazing.

Cast Brass—Brass is often rough-textured because it is usually cast in sand molds. Cast brass is used for water-valve housings and, with bronze alloy, for boat propellers. If you cut or machine cast brass or bronze—a brass alloy—it appears smooth and yellow or gold in color, depending on the specific alloy. Brass should not be ground. Soft metals such as brass load up, or clog, abrasive grinding wheels, quickly rendering them useless.

Cast brass is non-ferrous and non-magnetic. It can be welded with flux-coated brass rod and an oxyacetylene torch.

Cast Aluminum—This metal is usually rough, but because it's cast at a much lower temperature than cast iron, it can be cast in molds with a smoother finish. Cast aluminum is used for late-style lawn mower engines, motorcycle crankcases, intake manifolds for automobile engines and, more recently, automobile engine blocks and cylinder heads. As with brass, don't grind aluminum with a stone wheel. It also clogs the abrasive. Cast aluminum is non-ferrous. It can be welded with TIG, and gas, and brazed with aluminum filler. Cast aluminum can also be arc-welded.

Sheet Aluminum—This metal is smooth and shiny. It can even be polished to a mirror finish. Sheet aluminum comes in thicknesses as thin as kitchen foil or as thick as 2" or 3" plates. Sheet aluminum is used for screen-door frames, airplane wings, race car chassis, lawn furniture, siding on buildings, and many other common applications. Sheet aluminum is non-ferrous. Certain alloys can be welded with gas or TIG. You can also arc-weld some sheet aluminum.

Titanium—This metal looks similar to stainless steel, but it's much shinier when welded or filed. Although relatively light, it is very strong. Titanium-alloy forgings, tubes and sheets are used in aircraft and race car construction. It is very expensive. Titanium doesn't give off sparks when ground with an abrasive wheel. A ferrous metal,

This is a paint marker pen for marking and writing on metal. Buy one and use it to mark scraps of metal for future ease of identification.

titanium can be TIG-welded, but shielding is critical.

Identification Markings

Most metal alloys have identification markings. For example, chrome-moly steel is usually marked "4130 Cond. N," for 4130 steel, normalized condition. The condition of metal indicates how its temper, or hardness, was achieved. For instance, it may be in an as-fabricated condition, work-hardened or heat-treated. The same applies to tubing and sheet chrome-moly steel.

Sheet aluminum is usually marked to indicate the basic alloy and its conditions. Typical examples of aluminum alloys are 2024-T4, 3003-H14 and 6061-0. Refer to the chart of aluminum alloys, page 124 for a more detailed explanation.

Sheet stainless steel is marked "301, 308, 316 or 347," for example, depending on the specific alloy. The markings are similar to those for sheet aluminum except that STAINLESS may also be printed on the sheet.

For an indepth discussion of popular steel and aluminum alloys, read HPBooks' *Metal Fabricator's Handbook* (see page 154 for ordering). Not only are important properties of metals discussed, but the specifics of each of the popular alloys are covered.

Permanent Metal Markers—In my own metal shop, I keep a paint-type metal marker handy to mark cut-off pieces of metal so I can identify them later. I mark the cut-off piece with the same identification that was used on the full sheet, such as 6061-0.-90" or 4130-N-.025" and so forth.

Ferrous or Non-Ferrous?—I also keep a small magnetic screwdriver in my shirt pocket for quick

TEMPERATURE OF WELDING FUELS

Fuel	Air °F	w/Oxygen °F
Acetylene (C_2H_2)	4,800°F	6,300°F
Hydrogen (H_2)	4,000°F	5,400°F
Propane (CaHa)	3,800°F	5,300°F
Butane	3,900°F	5,400°F
Mapp Gas	2,680°F	5,300°F
Natural Gas (CH_4+H_2)	3,800°F	5,025°F

COLOR OF STEEL AT VARIOUS TEMPERATURES
in Fahrenheit (Centigrade)

Faint Red	900 (482)
Blood Red	1,050 (566)
Dark Cherry Red	1,075 (579)
Medium Cherry Red	1,250 (677)
Cherry Red	1,375 (746)
Bright Red & Scaling	1,550 (843)
Salmon and Scaling	1,650 (899)
Orange	1,725 (941)
Lemon	1,825 (996)
Light Yellow	1,975 (1079)
White	2,200 (1204)
Dazzling White	2,350 (1288)

Colors as viewed in medium light, not bright sunlight.

COLORS FOR TEMPERING CARBON STEEL

Color	Metal Temp
Pale Yellow	428 (220)
Straw	446 (230)
Gold Yellow	469 (243)
Brown	491 (255)
Brown & Purple	509 (265)
Purple	531 (277)
Dark Blue	550 (287)
Bright Blue	567 (297)
Pale Blue	610 (321)

Carbon steel assumes colors when heated.

identification of metals. Why? Because I cannot always visually identify the metal I'm working with. As already discussed, a magnet attracts ferrous metals—those containing iron. Magnets are not attracted to non-ferrous metals—metals that contain little or no iron, such as aluminum.

So, I touch my magnetic pocket screwdriver to the metal I want to weld. If it sticks, the metal has iron content. If the magnet doesn't stick, the metal has little or no iron content. But this is only a preliminary test. Some non-magnetic metals, such as stainless steel, cannot be fusion-welded to other non-magnetic metals, such as aluminum or magnesium. Such dissimilar metals must be joined by brazing or soldering.

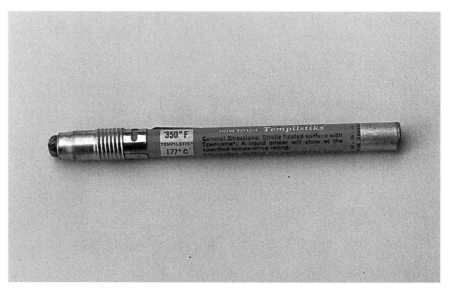

This is a heat-measuring Tempil stick. You can see 350°F printed near the tip. Mark the chalk-like stick on metal and the mark will melt at 350°F plus or minus 2°.

Mistakes Will Happen

Of course, there's more to identifying metals than determining if they're ferrous or non-ferrous.

Even though cast iron and mild steel are both magnetic and ferrous, the two cannot be joined using conventional welding methods. Special care must be taken to choose the proper welding process and welding rod.

Several years ago I was in a big rush to weld several aluminum stove hoods for a local restaurant. My supply of welding rod included short pieces that had been put back in the holder by a previous welder. By mistake, a piece of 316 stainless rod had been put back in the aluminum rod holder.

I did not notice the heavier weight of the stainless rod compared to the aluminum rod, and I proceeded to weld at least 12 inches of .090" aluminum stove hood seam with the stainless rod!

The dissimilar metal weld did not fuse together. Stainless steel will not mix with aluminum. I had to cut the stainless rod out of the seam and make a patch to cover my mistake.

Stainless steel rod is not magnetic and aluminum rod is not magnetic, but the aluminum rod is only one-third the weight of stainless rod. I made a mistake!

WELDING TEMPERATURE

With the frozen state of water in mind, consider the same condition for a piece of steel plate. When the steel is "frozen," below about 2,700°F, its melting point, it is solid. When heated above that temperature, steel changes to a liquid. And if you heat it to a temperature considerably above the melting point, steel can boil and vaporize into the atmosphere. You obviously shouldn't vaporize a weld bead, so keep the weld-puddle temperature below the boiling point.

Overheat at First

Most beginning welders don't heat the base metal enough. They fail to get it to the melting point and keep it there. I usually tell the beginner to go ahead and burn up, or overheat a few practice pieces. Doing this gives the beginner a feel for temperature control. Melting temperatures vary considerably from metal to metal. If you can master temperature control, you have a head start on welding successfully.

Temperature Control

If you can learn to correctly control the temperature of the weld puddle—the molten pool that forms the weld bead—you can do a good job of welding. To control the weld puddle, you must learn how to judge and control the temperature of the metal you're welding.

Getting back to the water example, water will pour and flow at room temperature. However, if you lower its temperature to below its melting point, or 32°F (0°C), it becomes a solid—ice. You can then handle it just like a block of steel; saw it, drill it or sit it on the freezer shelf without the need to contain it.

If you put this block of solid water in a pan and heat it to 212°F (100°C), it begins to boil and vaporize as it changes from a solid, to a liquid, then to a gas. Steam boils off and escapes into the atmosphere. Water, therefore, has a freezing point, a melting point and a boiling point.

Once you've mastered temperature control, the next thing to do is to become familiar with which metals can be welded and by which methods. The accompanying charts give the melting points of metals that can be welded, brazed, and soldered.

Make photocopies—you have the author's and publisher's permission—of the five charts in this chapter. Post them in your welding shop for quick reference. You'll be amazed how helpful they'll be.

WHY CAN'T I WELD WITH PROPANE?

The chart on the previous page shows 3,800°F flame temperature. The small bottle of gas I purchased from the hardware store is labeled 2,500°F (1371°C).

It CAN be used to weld and cut with, but the efficiency is below what you need to actually weld with. Pure oxygen and pure acetylene mixed together in a concentrated flame will give 6,300°F, a welding flame.

Propane and air are not as efficient, because air contains nitrogen, which is non-combustible.

Chapter 2
AN OVERVIEW OF WELDING, BRAZING AND CUTTING

Oxyacetylene fuel is used in a torch handle to gas weld on aluminum bracket. Flux cored filler rod makes this process very easy to do.

Historical records tell us that iron, gold, silver and bronze were commonly used as far back as 6,000 years ago. Biblical records indicate that metal swords existed even in the time of Adam and Eve. However, the significant joining of metal became common practice with the invention of the forge, which superheated fire by blowing air into a bed of red-hot coals.

Oxygen mixed with acetylene gas and burned in a blowpipe was first used for metal fusion welding at the turn of the twentieth century (1900). The invention of oxyacetylene welding did a lot to promote the aircraft industry in World War I.

Arc welding was in its infancy in the early 1900s. My earliest memories as a child were of watching an arc welder at work and getting my eyes burned from the arc. That was in 1938. I was 3 years old then. But the first arc welding rods were merely bare steel rods.

Oxyacetylene really became a well-developed art in building the WWII steel tube fuselage airplanes. The two gases, oxygen and acetylene, did not change but the regulators, hoses, and welding torches improved. My first torch was a 1950 vintage Victor Aircraft torch, and it did a very good job of building go-karts for me from 1958–1962.

TIG (Tungsten Inert Gas) welding was developed during WWII to improve the ability to weld aluminum parts for airplanes. It is also referred to as the trademark name Heli-Arc welding. This type of welding began to improve in the 1970s and 1980s when it utilized inverter technology. Inverter technology allows much smaller and lighter transformers than with straight AC or DC welding equipment.

In 1970, a typical AC-DC Heli-Arc welding machine weighed 400 to 500 pounds. Today a machine with even greater capabilities will weigh less than 50 pounds because of the use of inverters, rectifiers and electronic circuitry.

AN OVERVIEW OF WELDING PROCESSES

In this chapter, I will provide an overview of the 10 most common methods of welding, brazing, and soldering metals together. This book is written to be a handbook for the beginning welder and as a source book for the more experienced welder. Schools and factory welding departments will find that memorizing pages of welding hints and tips is not necessary if you have Welder's Handbook handy. The processes described in this chapter are: (1) Gas (oxyacetylene) welding and heating, (2) torch (oxyacetylene) cutting, (3) gas brazing and soldering, (4) arc (stick) welding, (5) MIG (wire feed) welding, (6) TIG (Heli-Arc) welding, (7) plasma cutting, (8) spot welding and (9) cold welding with epoxies.

Oxyacetylene Welding and Heating

The heat produced by an oxyacetylene (gas) welding or heating torch is 6,300°F (3,482°C), which is very adequate for most welding projects. The heat control techniques used in gas welding are very similar to the heat-control techniques required in most other kinds of welding, therefore, gas welding is usually taught first in most formal welding schools.

The size of the torch is directly related to the capacity or thickness of the metal it will weld or heat-form. A small torch set up for welding aircraft parts would usually weld steel no thicker than 3/16", .1875" or 4.7mm. A large torch set up for use in oil field work might be capable of fusion welding steel up to 1" (25mm) thick.

If I could only afford one welding machine for my welding

shop, I would pick a gas welding setup. Next would likely be a wire-feed (MIG) unit. Current, modern welding equipment is readily added to by use of modular units that have specific purposes.

Oxyacetylene Brazing & Soldering

Any torch that will weld can also be used for brazing and for soldering. An oil-field-sized torch that will fusion-weld 1/2" thick steel can be used to braze and solder by reducing the gas and oxygen pressures at the regulators, by using a small tip, and by adjusting the flame to a very soft, low heat output flame. However, if accuracy is important in your project, such as jewelry making, a torch setup that is suited to the required heat output will be the best.

Flux—When torch brazing and soldering, it is necessary to use flux to clean the parts while heating. Flux is a chemical powder or paste that cleans the base metal and protects it from atmospheric contamination during soldering or brazing (see glossary for a more detailed explanation). This may be done by using flux-filled or flux-coated rods and solders. Flux is a cleaning agent that comes in many forms, each suited to the base metal and the joining process used.

When preparation is done properly, brazing and soldering are the easiest of all metal heat-joining processes. First, the metal is prepared, cleaned and fitted for joining. Next, the appropriate amount of heat is applied to the parts with the torch flame, and then the filler metal and flux are applied. When the seam or joint is completed, the heat is removed by taking the torch flame away from the part. When the part cools, water is used to wash off the flux. That's it. Most people can learn to braze and solder in just a few minutes. But preparation is the key to success in brazing and soldering.

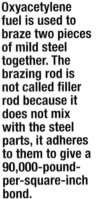

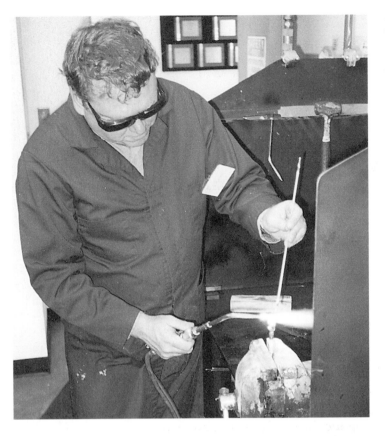

Oxyacetylene fuel is used to braze two pieces of mild steel together. The brazing rod is not called filler rod because it does not mix with the steel parts, it adheres to them to give a 90,000-pound-per-square-inch bond.

Oxygen and acetylene are mixed and burned by this cutting torch to cut through a thick steel bar in just a few seconds.

Oxy-Flame Cutting

Only certain metals can be oxidized (rusted) and therefore only those metals can be cut with a cutting torch. Stainless steel and aluminum do not rust, and therefore the oxygen from a cutting torch will not effectively cut those metals even though the oxy-fuel cutting torch has enough heat to melt them.

To effectively demonstrate the oxygen-oxidizing process used in torch cutting, I describe a trick of

cutting steel plate with oxygen only on page 78. To better understand the cutting torch process, read that special sidebar. In addition to plasma and laser beam cutting processes described in this book, there are several other cutting processes, including air-arc gouging, magnesium arc gouging, abrasive wheel cutting and, of course, saw blade cutting.

Oxyacetylene cutting requires more practice to

Arc welding is being practiced by this welding student at Allan Hancock College. A stick of flux covered 1/8" diameter steel rod/wire is called a consumable electrode, because it provides the spark heat for the weld, then becomes the filler metal as it melts into the part being welded.

perfect than other welding processes, but it is a very handy skill to have when you are building a metal project. Auto dismantlers also make good use of oxygen-propane cutting torches when salvaging reusable parts for cars and trucks.

Arc or Stick Welding

For many years, the accepted method for putting big things together was by arc welding, also called "stick welding" in the trade. The term stick welding was used because the coated welding rod was a stick of wire about 12 inches long, coated with flux to provide for strong, defect-free welds. Welders also called it stick welding to differentiate from wire-feed MIG welding.

Many of the world's tallest skyscraper buildings were put together with hot rivets and strengthened by arc welding. Most of the nuclear powerplants in the world were welded with low hydrogen welding rod and arc welding machines.

Arc welding is somewhat difficult to learn, but a small shop will find that many things can be welded with the average arc welding machine. A small 220 amp/220 volt buzz box arc welder can build 2- and 4-wheel trailers very adequately.

SMAW—Shield Metal Arc Welding (SMAW) is the oldest and simplest form of electric welding. Many small shops have a buzz box arc welder in the corner somewhere. Since 1940, many hot rods, race cars, utility trailers, farm repairs and auto repairs were successfully done with transformer-powered arc welders.

The stick electrode is usually a piece of steel, stainless steel, or aluminum rod that is coated with a flux of a clay-like mixture of fluorides, carbonates, oxides, metal alloys and binders to provide a gas shield for the weld puddle. When the rod is used in welding, the coating cleans and provides a protective cover for the hot weld bead to protect it until it cools below the critical point where the atmosphere could degrade the weld bead.

Arc Cutting

An arc welding machine can be used to sever steel, cast iron, aluminum, stainless steel and even the most exotic metals. Two primary methods can be used. The industry standard for many years has been the air-arc that uses air pressure combined with the 6,000°F to 10,000°F arc temperature to blow away the molten metal in a somewhat wide *kerf* (width of the cut). A 3/16" air-arc rod will usually cut a 1/2" wide kerf, and the kerf is usually more jagged than when the metal is cut with an oxyacetylene torch.

Because of the way the air-arc works, most welders call it the *air-arc gouging* process. The primary use of this process is to cut out defective welds on thick sections, usually over 1/2" thick. The arc is started as in arc welding and as soon as a puddle of molten metal is established, the arc rod is laid nearly parallel with the metal surface, and high pressure air is triggered to blow through the hollow arc rod, blowing away metal as soon as the 10,000°F arc makes the puddle. It is a quick-and-dirty method of metal removal.

And it requires a special electrode holder that has provisions for forcing 90 to 150 pounds of air pressure through the arc rod. A fireproof blanket must be used to catch the sparks blown out of the kerf. Another product of air-arc gouging is noise, caused by the high pressure air. Ear protection is a requirement when using the air-arc process.

Magnesium Arc Rod—A somewhat newer process for cutting and gouging metal is the magnesium-based arc rod produced by specialty manufacturers. The best feature of this cutting rod is that it requires no special equipment such as air pressure and special rod holders. It can be used with any standard stick electrode holder.

This rod can be kept in a toolbox and used for cutting off rusty bolts and nuts, severing metal where accessibility is a problem, and where oxyacetylene cutting would be impossible, such as when cutting aluminum plate and castings, stainless steel, and even cast iron. The method of use is similar to air-arc gouging. The arc is struck and then the rod is laid nearly parallel to the surface and pushed forward to extrude the metal from the kerf. Several passes must be made to cut through 1/2" metal plates. This cutting/gouging rod is economical in terms of time and money. In a particular case where NASA required an exotic metal to be cut into blocks for machining, the cut for each block took over one hour and expended a $100 bandsaw blade. The same cut could be done with Cronatron 1100 rod in 5 minutes at a cost of $5 for the cutting rod. See www.cronatronwelding.com.

MIG Welding

Wire-feed welding, officially known as GMAW (gas metal arc welding), has really become popular since 1985. It is possible to drive to your local hardware store, buy a $300 MIG welder, take it home and plug it into a 110V outlet in your shop and immediately begin welding.

But, as with most things in life, one machine just will not do everything. The least expensive MIG welder must use flux-cored filler wire, and it is limited to 3/16" and thinner metals, and it produces gobs of smoke and spatter. But when you are building a barbecue grill or making shop equipment, or ornamental iron, you can make the least expensive machines work for you.

On the opposite side of the picture, robotic MIG welders that assemble new car bodies and components cost tens of thousands of dollars and make welds that are so good that the parts appear to be one piece, with little or no evidence of welding.

Although MIG welding is the easiest kind of welding to learn at first, it is also one of the more difficult processes to master. Even in the cutting-edge race car fabrication business, many people find that accurate weld seams are hard to master because of at least three reasons:

• You can't see the puddle easily because the nozzle of the MIG gun blocks your view. A partial solution to this is to look at the weld puddle from the side rather than from behind the MIG gun.

• The smoke and sparks that are a result of the flux-cored MIG process make seeing the puddle very hard to do. A partial solution to this problem would be to use inert gas and solid wire to reduce the smoke.

• The weld begins when you squeeze the MIG gun trigger and continues until you release the trigger. You are committed to weld, ready or not, so make sure you practice first.

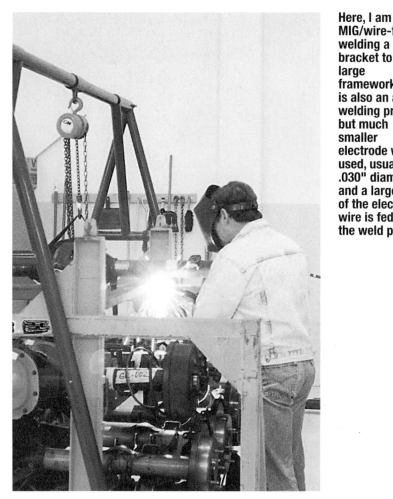

Here, I am MIG/wire-feed welding a bracket to a large framework. MIG is also an arc welding process, but much smaller electrode wire is used, usually .030" diameter, and a large roll of the electrode wire is fed into the weld puddle.

TIG Welding

This is the elite of all welding processes. You can operate a TIG GTAW (gas tungsten arc welding) torch while dressed in a white tuxedo. And the weld you do will be neater, probably stronger, and easier to do than with all other welding processes that require fusion of two or more pieces of metal.

The least expensive TIG welders make an inert gas protected weld puddle, but the heat is not adjustable while welding. You are committed to whatever heat setting you decided on before you started, or you must stop and readjust the heat. Test welds done before welding the valuable part will partly solve this problem.

The best but most expensive TIG-welding machine can be programmed to make nearly perfect welds with only moderate input from the operator/welder. Pulsed TIG can give you short or long applications of heat with cooling times between application of heat so that your dipping the filler metal into the molten puddle will coincide with the pulses of heat from the machine.

Foot-pedal-operated TIG machines allow the welder to start a puddle, hold the puddle while the

In this photo, I am TIG-welding a steel tube fuselage repair on an experimental airplane. Bare steel welding rod is being fed into the weld puddle just as in gas welding. The welding torch heat is being controlled with a foot pedal amp control. Photo Jim Holder.

Plasma Cutting is a fast way to reproduce large number of similar parts for production welding and fabrication shops, This cutter is being operated from a laptop computer with a servo head that has been programmed by a high school student in his dad's welding shop.

heat spreads into the part, hold the puddle while you reposition yourself or the filler rod, and they allow for a cool-down of the puddle before shutting off. Automatic TIG welders can be programmed to make perfect welds on things such as the fuel system plumbing on the space shuttle and in nuclear power plants.

An ideal feature of many of the top-end TIG machines is that they can be used to stick weld, with the appropriate additions of auxiliary equipment.

Plasma Cutting

This is another welding/cutting process that became very popular in the late 1980s, to the point that almost every sheet metal fabricating shop and metal fabricating shop can afford one. Plasma cutters are extremely easy to operate. Anyone can be given one minute of instruction and then can make a cut in any metal. Setting up the machine and replacing the consumable nozzles and electrodes takes just a little more instruction, but still, any shop helper can do it.

Cutting stainless steel sheet and tubing is one of the more difficult things to do in a metal fabrication shop, but a plasma cutter can just as easily cut stainless steel as any other metal, including mild steel. All metals can be cut with a plasma cutter, including aluminum, brass, copper, titanium. When cutting over 1/2", other cutting methods should be considered.

Spot Welding, Laser Cutting, EBW

Spot Welding—Spot welding is within the capabilities of any metal fabrication shop. Mild steel and low carbon steel can be spot welded with the least expensive equipment. Aluminum spot welding is very common in aerospace and in mass-produced commercial equipment. A body shop spot welder can be operated from most MIG, TIG and stick welding power supplies.

Aluminum spot welding machines are much more complex. The average spot welder for aluminum is twice as large as a refrigerator and new costs are around $100,000. Aluminum requires much more precision to weld than does steel.

Laser Beam Cutting—Laser beam cutting is a more popular cutting method. Many fabrication shops contract with specialty shops to laser cut metal parts that are ready to use as cut. Lasers can easily cut through most metals, up to 1" thick. Thinner materials, .025" through .375" are ready-made for super cuts by laser beam. Laser beam cutting leaves no slag on the backside of the cut as other methods do. One LBC specialty shop hands out key rings with their logo cut in .08" stainless steel, and the key ring is as smooth on the back side as it is on the top side. Look in the yellow pages of your phone book for laser cutting.

EBW—Electron beam welding (EBW) is not a process for small fabrication shops, and is mentioned here only to familiarize you with this process. As the name implies, EBW uses a thin electron beam to fusion-weld two or more parts together when other welding methods would not be possible. Large manufacturers use this process to join automotive transmission gears to a common shaft, missile component manufacturers use the process to join intricate parts that must be made in pieces and then joined as if the final product were one strong, complicated part. Most EBW is done in vacuum chambers with inert gas to prevent atmospheric contamination. Specialty fabricators in most large cities can provide EBW services in small lots of one or more parts projects. Laser beam welding (LBW) is similar to EBW, and can be performed by specialty shops.

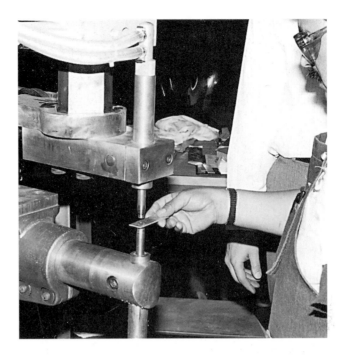

This large, expensive spot welding machine is being tested for spot welding aluminum. Once they are properly adjusted, spot welding machines are very easy to operate.

Natural gas, super-heated by injecting large amounts of air into the flame, is used in this ornamental ironworks oven to heat steel bars for forming into fireplace tools.

METALS & WELDING PROCESSES

The following chart will help you determine which filler materials are available for which metals, and the welding process that can be used to weld each respective metal.

PROCESS

Material	*Thickness	Arc Weld.	MIG Weld.	TIG Weld.	Spot Weld.	Gas Weld.	Electron Beam	Oxyacet. Brazing	Lead Solder
Carbon Steel	S	X	X	X	X	X	X	X	X
	I	X	X	X	X	X	X	X	X
	M	X	X	—	—	X	X	X	—
	T	X	X	—	—	X	X	—	—
Low Alloy Steel	S	X	X	X	X	X	X	X	X
	I	X	X	X	X	X	X	X	X
	M	X	X	—	—	—	X	—	—
	T	X	X	—	—	—	X	—	—
Stainless Steel	S	X	X	X	X	X	X	X	X
	I	X	X	X	X	—	X	X	X
	M	X	X	—	—	—	X	—	—
	T	X	X	—	—	—	X	—	—
Cast Iron	I	X	X	X	—	X	—	X	X
	M	X	X	—	—	X	—	X	—
	T	X	X	—	—	X	—	X	—
Aluminum	S	X	X	X	X	X	X	X	X
	I	X	X	X	X	X	X	X	X
	M	X	X	X	—	—	X	X	—
	T	X	X	—	—	—	X	—	—
Titanium	S	—	X	X	—	X	X	X	X
	I	—	X	X	—	—	X	—	—
	M	—	X	X	—	—	X	—	—
	T	—	X	—	—	—	X	—	—
Copper & Brass	S	—	X	X	—	X	X	X	X
	I	—	X	—	—	—	X	X	X
	M	—	X	—	—	—	X	X	—
	T	—	X	—	—	—	X	—	—
Magnesium	S	—	X	X	X	—	X	X	—
	I	—	X	X	X	—	X	X	—
	M	—	X	—	—	—	X	—	—
	T	—	X	—	—	—	X	—	—

*
S=sheet up to 3mm, 1/8".
I= 3–6mm, 1/8-1/4".
M= medium, 6–19mm, 1/4–3/4".
T=thick, 19mm, 3/4" and up.

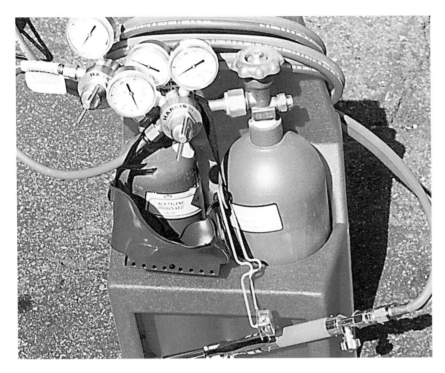

Chapter 3

WELDING EQUIPMENT

The first welding setup for any shop should include a portable gas welding (oxyacetylene) torch, carrier and bottles such as this Harris torch.

For welding shops ready to upgrade their welding equipment, the best process is to shop around for the best equipment that suits their purpose. In almost every instance, the smartest thing to do is to try the equipment before you buy it unless it is a welder with known capabilities. If your friend has a welder made by a specific company and he is happy with it, then your choice of welder is narrowed down significantly.

But with steady improvements in technology, it is advisable to shop around and ask to be introduced to the latest and best the industry has to offer. Electronic welding helmets were not even invented fifteen years ago, and now they are on every welder's shopping list. Pulsed MIG and TIG welders offer better control of the weld, both for the novice welder and the professional welder. Inverter technology has really opened up the welding machine industry to provide portable welders that will do more than the big, heavy machines made in the 1960s and 1970s.

WHERE TO BUY

Local welding supply dealers usually sell at retail except for sales promotions. Large department stores and building supply stores usually have tool departments, and most of these tool departments have welding equipment for sale. Mail-order catalogs are other good sources for welding equipment. Most of the automotive and aircraft monthly magazines feature advertisements from reputable welding supply firms, and their ads provide good information about prices and features. Home improvement centers are also good sources of welders and welding supplies.

For occasional use, check with tool rental firms, and don't overlook used welding equipment listed in newspaper want ads. However, buying used welding equipment is about as risky as buying a used car or used lawn mower. In every case, trying the equipment before you buy is a very good practice.

GAS WELDING EQUIPMENT

For sure, a gas welding torch setup is strongly advised in any beginner's shop. Being able to heat, bend, form and torch-cut steel parts is essential. Even large, well-established fabrication shops can always benefit from a lightweight, portable gas welding and cutting setup. The most important point to decide on when purchasing gas welding equipment is whether you will be working with very thin material— 0.49" and thinner —or with thicker material, 0.50" to .250" thick and thicker. For thin material, choose an aircraft-type torch and possibly add a jeweler's torch to your toolbox. If your projects will be race cars, trailer or farm equipment, a standard-sized torch is recommended.

ARC WELDING EQUIPMENT

Several different brands of stick electrode arc welding equipment are available. Usually, these 220 volt, 225 amp to 250 amp buzz box welders are priced about the same as a small portable oxyacetylene torch setup.

The pros and cons of buzz-box welders are similar to the same for MIG-wire-feed welders:

Most initial gas welding sets include single stage regulators like the one shown here in cut-away. Tightening the top screw will compress the large spring that acts on the diaphragm to provide pressure to the torch. Victor Equip. Co.

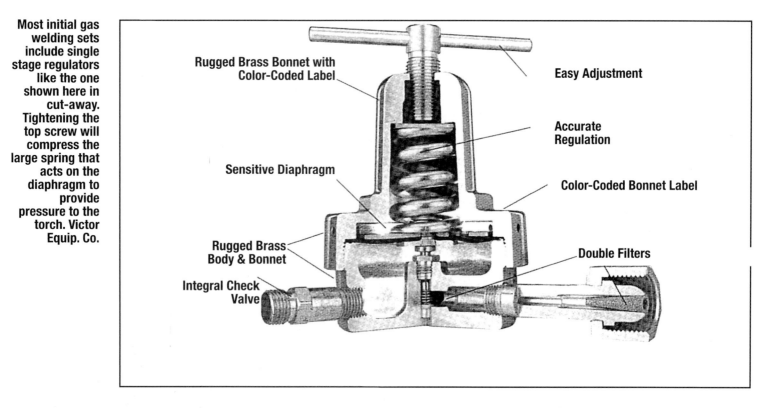

Rugged Brass Bonnet with Color-Coded Label

Easy Adjustment

Accurate Regulation

Sensitive Diaphragm

Color-Coded Bonnet Label

Rugged Brass Body & Bonnet

Double Filters

Integral Check Valve

This Lincoln Electric AC-225 "Buzz Box" arc welder will last a lifetime and it sells for about $300. It will stick weld and cut metals with magnesium cutting electrodes. It will build trailers and similar equipment, and even dirt track race cars.

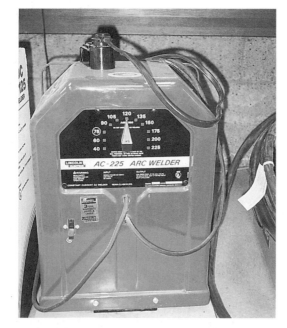

•Buzz-box welders are only useable with coated stick electrodes. Stick electrodes have a shorter useful life than bare welding rod and welding wire because moisture affects the flux coating and reduces the arc rod's welding ability. Old, wet stick welding rod is almost useless and must be disposed of.

•Buzz-box welders are not production welding machines because they have a duty cycle of 10% to 40%, meaning that they must rest 6 to 9 minutes in every 10 minutes of welding.

Positive aspects of buzz-box welders are:
•You can purchase coated stick welding rod in small (1 to 5 lb.) packages, and that way you have fresh rod for each project. And you can purchase small amounts of several kinds of welding rod.

• Even though 20% duty cycle means that you can weld 2 minutes, and that you must let the welder cool for 8 minutes before resuming, that is not a serious problem for most small projects. Changing used-up rods, tack welding and moving around to the other side will average out the 20% factor.

Larger (up to 500 amp) arc welders are available, but most shop welding jobs are done with 1/8" rod at 130 amps or less, usually at 100 amps.

Welding schools, fabrication shops, production welding shops and construction shops are users of the higher amp arc welding machine.

Another feature of large arc welding machines is that they are usually able to be switched from AC to DC positive and DC negative polarity for special welding jobs.

MIG WELDING EQUIPMENT

By far the most popular small shop welder is the 110-Volt MIG/wire feed machine. With a 5-pound roll of flux-cored wire installed, these machines are all ready to plug into house current and weld

This photo is of the Gibbs Racing shop, showing the many NASCAR racers that are being prepared for racing. Note the cleanliness of the work area. The cleaner and better lit the shop, the better your welding will be. Courtesy Lincoln Electric Co.

immediately. There are two drawbacks to the 110-Volt MIG welders:

• Flux-cored wire feed welding produces lots of smoke and spatter, and the spatter would be objectionable in appearance in most welds.

• The bottom-of-the-line MIG welders cannot be converted to gas and they cannot be converted to weld aluminum. You can only weld steel with them.

Some of the positive aspects of portable MIG welders are:
• They are almost always ready to weld. You simply adjust the voltage knob, adjust the wire speed knob, turn on the power switch, and you are ready to tack weld or to weld.

• It is possible to weld for several hours without ever changing the electrode (wire roll).

• Higher capacity (and higher cost) MIG welders are so trouble-free that they are used in robots to weld automobiles in auto manufac-turing.

• Lower- to medium-priced MIG welders are available that can be switched to gas, and the polarity can also be switched for gas and aluminum welding.

TIG WELDING EQUIPMENT

TIG welders are available in more configurations than most other welders. All TIG welders must be supported with argon gas or other shielding gas, but as one manufacturer states in its welding booklet,

The smallest MIG welding machine that is practical for most small welding shops, is this one that is rated at 175 amps. It can weld steel and with liner changes, it will weld aluminum. This one is set up to weld steel with gas, not flux core.

This small, 110 volt, 95 amp buzz box welder only cost $120 at the local department store, but it is severely limited in what it will weld, and it requires special, small diameter welding rod that has magnesium to make it easier to start the weld bead. Don't waste your money on this size machine.

Another important tool that is good to have in a well equipped shop is this manual tubing bender that is made by Pro-Tools. Jason Woods of Alamogordo, NM, is shown here demonstrating the lever used to bend the tubing up to 2.25" diameter and with .125" wall thickness.

This is a common MIG welding machine that is set up for gas, therefore the gauges and regulators are behind the welder. Notice that it is only a 125-amp, 110-volt machine, capable of only welding tailpipe and body and fender thickness. Spend a little more money and buy at least a 175 amp welder.

John Gilsdorf of Alamogordo, NM, is restoring and customizing this 1956 Ford Parklane 2-door station wagon. It will feature this 1969 Ford Thunderbird 429 cid engine that has 11: to 1 compression and backed up to a C-6 automatic transmission. His shop is dedicated to restoring cars.

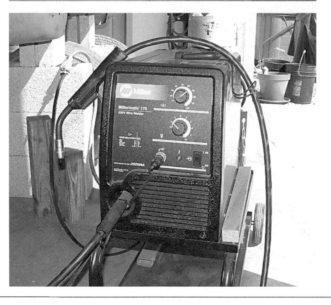

John Gilsdorf's choice of a MIG machine is this Miller Electric 175 amp unit that can also weld brackets and sheet metal. It is gas supported rather than being flux-cored welding wire.

I cut my TIG welding teeth on a machine just like this one. It was made in late 1969 or 1970, but it still does high-quality welding. Look for a machine like this at bankruptcy sales, school surplus sales.

TIG welding can be done with a regulated argon flow and 24 volts from two car batteries wired in series! Safe to say, there would be no current control in a setup like that, so you need to decide which welding setup suits your needs.

Scratch-start TIG torches do a pretty good job if you are welding thin to medium thickness steel tubing and steel sheet. There are several brands of DC-only TIG welders that use scratch start and pulse current to weld steel. You cannot weld aluminum with a DC-only welder. Before buying one of these welders, try it out, or at least ask for a satisfaction-or-your-money-back guarantee. They do excellent work on 4130 steel and mild steel. They cost about 2 1/2 times more than buzz box welders.

Foot-pedal-operated, variable remote current control AC/DC, high frequency TIG welding machines are a dream to operate. The bottom-of-the-line, full-feature TIG welder sells for about 4 to 5 times the cost of a bottom-of-the-line buzz box AC welder or DC-only MIG welder, but they will do many kinds of fusion welding on most materials, and the welds are aircraft-quality welds.

The top-of-the-line TIG welders feature pulsed, square wave current, pre-flow adjustments, postweld timer, and repeatable weld programs. Many of them have digital readouts, and will cost about as much as a small economy car. But they make an average welder look like a welding talent!

PLASMA CUTTING EQUIPMENT

Good, low-priced plasma cutters can be found in many mail-order ads, in car magazines, in aircraft magazine ads, and in large department stores, as well as in local welding supply outlets.

The best thing about plasma cutters is that they will cut stainless steel as easily as they cut mild steel. If you have ever tried to hacksaw or bandsaw cut a sheet of .080" thick stainless steel sheet, you know that it will ruin your blade in just a couple of inches of cut. With the plasma arc cutter, you can cut a 4' x 8' sheet of stainless steel in half in about 2 minutes, and do it again and again before you have to replace the consumable electrode in the torch handle.

Plasma cutters come in many sizes and prices, just like other welding equipment. The bottom-of-the-chart plasma cutting machine costs $1,000, and a top-of-the-line, BIG plasma cutter costs $5,000 or more. The differences are in what the machine will do. They all will cut all metals—steel, aluminum, copper, stainless steel, brass, and titanium—but some plasma cutters do the job better and cleaner than others do.

Even the higher-priced plasma cutters have duty cycles. If you cut for 10 minutes without stopping, their "fire" will die out and not come back on until the unit cools for 10 minutes. If possible, try the plasma cutter before you buy it, or get a satisfaction-or-your-refund guarantee.

SPOT WELDING EQUIPMENT

These sheet metal assembly welders come in two classifications. The transformer-type will spot weld steel, galvanized steel and, sometimes, stainless steel. The second type will spot weld aluminum, magnesium and titanium by special programmed weld sequences. These electronics-controlled, air- or hydraulic-operated capacitive discharge welders are very large and very expensive. Never, under any circumstances, should you buy an aluminum spot welder unless the sales contract provides for setup

Your shop will be well equipped when you have a plasma cutter like this one from HTP. It will cut any metal that is electrically conductive, steel, aluminum, copper, brass an more. It runs off 110 volts and 60 psi shop air.

and hundreds of samples performed and certified to prove that the aluminum spot welder will perform as you want it to.

One of the best methods for deciding on which welding equipment to buy is to visit several shops, welding shops, airplane or race car fabrication shops or ornamental iron shops, and ask the people there for their recommendations. Then make up your own mind and buy or rent the equipment that best suits you. In Chapter 8 through Chapter 17, this book will tell and teach you how to use the equipment.

WELDING AREA EQUIPMENT

There are a number of items you should have in your welding shop that will make your job a lot easier. Some are absolutely essential, others are optional. I'm talking about equipment, and not safety gear. Helmets, gloves, clothing, etc., will be covered in the Chapter 5.

Chair—Get a lightweight, comfortable chair. I prefer a chair like that used by a draftsman—one that can be raised or lowered to suit the work height. The chair should have rollers so it can be moved without having to pick it up. Don't spend a lot of money. Check a used-furniture store or a flea market first.

Table & Clamps—Build a metal-top welding table. The table top should be steel—not aluminum or wood—at least 1/4" thick and 2' square. Make the top larger if you plan many big, heavy welding jobs. A good all-purpose welding table uses a 3/8" thick, 3' x 4' steel plate for the top. The frame to hold the welding table can be anything sturdy enough.

Here John demonstrates a handy tool that he invented for pulling out small dents that are not accessible from the back side. He had MIG welded a plain cadmium-plated wood screw to the dented area and now is pulling the screw with a custom-made slide hammer puller that saved buying a $350 stud welder and puller.

The most economical way to heat large quantities of steel parts for ornamental iron projects is to use air-charged natural gas ovens like the one shown here. The heat output is over 3,000°F, but in large amounts of BTU's. Photo: Kummer Ornamental Iron.

Keep the tabletop rust-free, especially if you are using it for arc welding or TIG welding. Rust is a poor electrical conductor and will even contaminate gas welds. I power-sand the top of my weld table occasionally to expose shiny steel. Never paint the top of a weld table because it blocks current flow from the ground electrode to the workpiece.

Drill a 1/2" or larger hole in the back side of the table top or frame and put a bolt and nut there for a more secure ground connector attachment. See page 137 for a welding table design you can build yourself.

Keep an assortment of C-clamps, weights, metal clothespin clamps and holding fingers nearby to hold the pieces in position while you weld. Never ask anyone to hold a small part while you weld it. They won't like it when the sparks fly and the metal heats up!

Sandblaster—A great addition to any welding shop is a sandblaster. You can buy a small portable sandblaster that holds 50 pounds of sand for about the price of 3 tanks of gasoline. I used one to completely sandblast the frame of a 1956 Ford 1/2-ton truck I was restoring. The job took about 6 hours, but the result was a like-new frame that was ready to accept a coat of polyurethane paint.

A sandblaster is also handy for cleaning small parts and numerous projects. But it does take several minutes to get the sand ready. And after sandblasting, you'll need a bath to wash all the sand off.

If you buy a 4–8 cfm siphon-type sandblaster, you'll need at least a 1- or 2-hp air compressor. A 1/2-HP compressor doesn't have sufficient capacity. Pressure-type sandblasters work two or three times faster, but will cost up to ten times more. The siphon-type sandblaster will do a good job of cleaning metal for the occasional, small welding job. For more frequent welders, and more complex

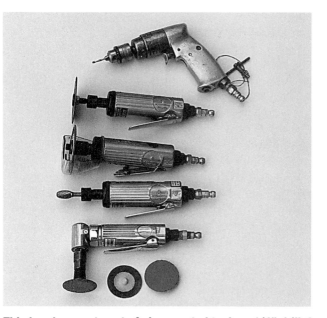

This handy assortment of air-operated tools, a 1/4" drill, 2 cut-off wheels, a die grinder and an angle sander, really make fitting and welding a lot easier. The average cost was $20 per tool.

welded assemblies, consider investing in a pressure-type sandblaster.

Metal Marker—A handy item to have around the welding shop is a metal marker. It's a tube of paint with a ball-point end that marks almost anything. It will mark smooth metal, rough metal, oily metal, wood, glass or even plastic. You'll find these markers at most welding supply shops.

This is a non-staged photo of the Chip Ganassi Racing Shop. This is how they really work and weld on their race cars! You can't have too much light in your weld area. The better you can see, the better your welds will be. I was lucky enough to catch a ride in a Ganassi Race Team vehicle at the Lincoln Welding Race car fabricating school a few years ago. Courtesy Lincoln Electric Co.

I use a metal marker for identifying scraps of metal that are suitable for future use. A tube costs about as much as a hamburger, but lasts about two years in normal use.

Never use a scribe for marking metal, except at cut lines. Scribe lines in metal act the same as perforations in a piece of paper–ready-made for tearing, breaking or cracking. If you do scribe-mark a piece of metal, the scribed area should be discarded because of the potential for cracks. Such a case would be marking a cut line.

Temperature Indicator—As discussed in Chapter 1, when welding or heating metal, it's crucial to get the metal to just the right temperature—no more. Excess heat will often ruin the metal. This can happen even when loosening rusted or otherwise stuck parts, too.

The way to check for exact temperatures is with a temperature-indicating crayon or paint. Temperature indicator is applied to the area to be heated or welded. When the indicated temperature is reached, the crayon or paint melts like wax on a hot surface.

Temperature-indicating crayons and paints come in more than 100 different temperatures, from 100°F (38°C) to 2,500°F (1371°C). Claimed accuracy is plus or minus 3°F. These and other types of temperature indicators by Omega Engineering and Tempil can be purchased at most welding-supply stores.

Chapter 4
WELDING RODS, WIRES AND FLUXES

Five-pound boxes of arc welding rods and brazing rods are available from local welding supply dealers. Know what rod is best for your welder and your project before you go shopping. Photo: SoCal Airgas.

Welding rod comes in grades of quality and strength, just like nuts and bolts do. Even though a rod is marketed and sold as 4130 rod, there are at least 6 grades of what the vendors call 4130.

You can buy at least 6 different grades of 1/4" fine-thread bolts, ranging from hardware store, no markings, no source, to aircraft-quality 1/4" fine-thread bolts with markings and even certification papers if you need them. The hardware store 1/4" bolt might get the job done if you are using it to repair lawn furniture, but you would not want to use hardware store bolts to hold the tail on your aerobatic airplane.

GRADES OF ROD

Let's look at the grades of welding rod, and then you be the judge of which ones you should stock in your welding shop. We will look at bare steel 36"-long rod as an example.

#1. Commercial grade, copper-coated mild steel. No brand, no source, probably made from remelted scrap steel, no known chemical content. Good for use as shop tie wire, nonstructural. Okay for use in welding rusty garden equipment and rusty tail pipes. Cost: $1.50 per pound.

#2. Commercial grade, copper-coated mild steel. Sold by a brand-name supplier. Probably the same metal as #1. Brittle welds, cost: $2.00 per pound.

#3. Commercial grade, copper-coated 4130 steel. Brand

name, but made from remelt, scrap steel, not certified. Cost: $5.00 per pound.

#4. Commercial grade, bare (no copper) 4130 steel. Brand name, certification papers say that the metal has 4130 material, specific carbon amounts. Cost: $10.00 per pound, minimum of 10- to 20-lb boxes.

#5. MC grade, bare 4130 rod. Brand name, certification papers specify new material, not scrap. Cost is $20.00 per pound in 10- to 20-lb. boxes.

#6. MC grade, vacuum melt, rolled, not die formed, certified 4130 and certified clean, no imbedded impurities, no copper, no oil, no rust, sealed containers, with desiccant dryers. Sells for $40.00 per pound in 6-lb minimum orders.

COPPER-COATED WELDING ROD

Welders use strips of copper to back-up their welds in their sheet metal because copper will insulate the weld and because it won't stick to the steel, stainless steel or aluminum that is being welded.

Copper strike strips are used to scratch-start TIG torches because copper will not wick up (flow) to the tungsten electrode.

Copper is never fusion-welded to steel because copper and steel won't mix.

Copper is used to coat welding wire to make it easier to draw the wire to 1/8", 1/16" or whatever diameter is desired.

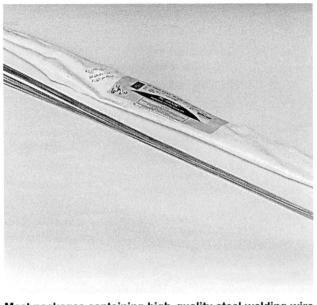

Most packages containing high-quality steel welding wire warn: DO NOT TOUCH WIRE WITH BARE HANDS OR UNCLEAN GLOVES. Welding rod must be kept clean and free of rust, so always keep it boxed to keep it free from contamination and dust. Courtesy United States Welding Corp. www.usweldingcorp.com

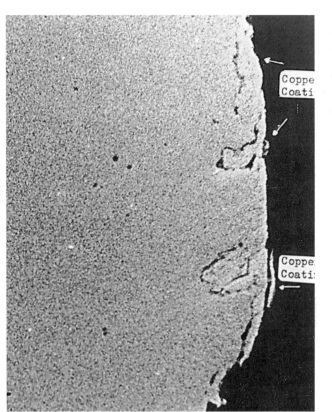

Cross-section of copper-coated steel welding rod magnified 3000 times. Copper coating looks foreign to the rest of the metal. It can do the same thing to your weld. Although it's OK to use for welding exhaust systems and metal furniture, don't use copper-coated rod to gas-weld expensive parts. Courtesy United States Welding Corporation. www.uswelding.com

CAUTION!

Copper fumes generated in welding are hazardous to your health. If you are welding with copper-coated welding rod or wire, be sure to evacuate the welding fumes with air suction equipment to prevent breathing copper fumes. Copper fumes accumulate in the body, which means that they can add up over a period of time. To be safe, just don't breathe copper fumes, or—better yet—don't use copper-coated welding products at all if you can help it.

Copper is applied to the rod or wire to lubricate the dies in the drawing or sizing equipment.

Copper is not applied to base welding rod to prevent rust. Copper-coated welding rod will rust if not protected from moisture in sealed containers. Even MIG wire in rolls should be kept in sealed containers to prevent moisture and dust contamination. Copper-coated MIG wire will flake off and sometimes clog the rollers, guides and cables and prevent proper operation of the MIG welders.

Copper-coated welding rod often causes cracks and bubbles in 4130 steel welds. Since copper does not fuse with steel, it flows into the grain of 4130 steel like brass does, and wedges the steel apart, causing cracks, and because copper-coated welding rod is usually made from inferior steel, it contributes contamination to the weld bead. If it is worth welding, take a little more time to do the job right, and use better-quality welding rod.

METALLURGICAL ADVANCES

In recent years, welding rod and consumable metallurgy has advanced almost as quickly as the electronics industry has advanced. When you are selecting a stock of welding rods, fluxes and wires, take a little extra time to search for the good stuff. In one specific example, I can spend $90 for the 4130 chrome-moly tubing to build an aircraft engine mount. The new 200-horsepower aircraft engine sells for $30,000. The primer and paint to paint the completed mount will cost $40. I need 1/2 pound of welding rod to weld the tubes together. I can either pay $1 for cheap rod or $25 for vacuum-melted, metallurgically controlled, certified welding rod that incidentally does a much better job. It makes more sense to use the $50-per-pound certified welding rod.

Similarly, there are solders that allow all metals to be joined, but these new high-tech solders cost 10 to 20 times as much per pound as the old 90-year-old technology solders do. But they get the job done, whereas the 90-year-old stuff doesn't. AWS

I enjoy telling welders that this load of scrap metal is what they make cheap welding rod out of, including old Chevys, old Fords , old Toyotas, with a little concrete rebar mixed in. The moral to this picture, is "don't buy cheap welding rod, it might be made out of this stuff."

(American Welding Society) classifications for welding generally indicate the correct rod for the material and welding process you are looking for. Consult the AWS charts whenever possible.

CHOOSING THE RIGHT ROD

Choosing the correct filler rod for welding a specific material is as important as knowing how to weld. You can buy scores of different types of welding rods at your local welding-supply outlet. There are hundreds of different types. So it's obvious that a welder needs some help in choosing the correct welding rod.

First, find out what kind(s) of metal you are welding. Then, decide which welding process is best for that metal. In many cases, you have several choices.

For instance, you may want to weld or braze a broken office chair. If the break occurred at other than a previously welded joint, you'd have at least four choices of how to fix it: gas-weld with oxyacetylene, arc-weld with small-diameter welding rod, TIG-weld it, or MIG-weld it.

In the case of repairing a break at an old weld, you'd first have to determine how it was welded, and then probably reweld it using the same method. Why? Some welds are not compatible with others. To determine the previous metal-joining method, make a visual inspection with a magnifying glass. An arc weld usually has some slag and spatter; a gas weld has flaking. TIG and wire-feed weld beads are clean. Brass braze is a gold or bronze color.

The easiest way to weld the broken chair, assuming it's a new break, would be to gas-weld it with either bare steel wire or copper-coated steel wire—if you have access to a gas welder.

Any of the other three choices would work.

To choose the correct welding rod, first determine the type of welding equipment available, then pick the correct welding rod. Welding rod choices are grouped according to the specific welding equipment.

In my home workshop, I have about ten different kinds of gas-welding and brazing rod, six kinds of arc-welding rod, and four or five kinds of solder. A commercial welding shop would have a similar assortment of filler materials. For TIG welding, you want at least ten different kinds and sizes of welding rod for welding common steel and aluminum alloys of various thicknesses.

In addition to being used as a filler material, welding rod has numerous other uses. For instance, you can use an arc welder and welding rod to build up a worn surface, such as a crankshaft journal. Or, welding rod can be used to add a hardened surface.

Lead Soldering w/Gas Welder

50/50 Wire Solder—This typically comes in spools and is used with brush-on flux to solder copper tubing with an oxyacetylene torch. Melting temperature is extremely low: 250–400° F.

Wire solder is available in several different alloys: 10/90, 15/85, 20/80, 40/60 and 63/37. In the recommended example, 50/50 indicates 50% tin and 50% lead composition by weight. High lead-

Moisture is a very detrimental thing to flux-coated stick welding rod. This Phoenix Dry Rod portable rod oven can be plugged into 110-volt power to keep your stick electrode dry and warm, right at the job.

NASCAR teams have plastic pipe tubes like these to store their bare welding rod in. Just go to the building supply store and buy some lengths of 1 1/4" plastic pipe and caps to make your own welding rod strorage tubes like these.

You can also go to your welding supply dealer and buy these plastic tubes to keep your stick electrodes in to keep them from getting wet or moist and ruining.

content solder such as 10/90, 15/85 and 20/80 are used for sealing brass auto radiators and filling seams in steel automobile bodies. These solders are used primarily with acid-base fluxes.

Inorganic Fluxes—These are the strongest, yet most corrosive. Because inorganic fluxes are corrosive, they should be used only where it's easy to remove them after soldering. They are usually made of salts such as ammonium chloride or zinc chloride dissolved in water. Use inorganic fluxes only when soldering copper or steel buckets, or small pieces that are easy to dip or wash clean.

Medium-Strength Organic Fluxes—These are used for soldering copper tubing and brass, as well as steel. Usually they are called acid fluxes because they are made from glutamic or stearic acid. Otay brand flux is called No. 5 soldering paste, and is for cleaning and fluxing all metals except aluminum, magnesium and stainless steel.

Rosin Flux Solders—These are the weakest type and must not be used for flame soldering. The base for rosin flux comes from pine-tree resin. It

activates when heated and deactivates as it cools. Use rosin flux only for soldering electrical and electronic components.

Aluminum Soldering w/Gas Welder

Welco 1509—This is a low-temperature (500°F/260°C) solder for joining aluminum, zinc, die-cast metal, copper, brass, stainless steel and other metals to each other, or to themselves. It requires a flux for cleaning and soldering. I use Welco 380 flux. This solder has a tensile strength—stress at which it breaks while under tension—of 29,000 psi. It comes in 1-lb. wire spools and is available in 1/16", 3/32" and 1/8" sizes. Welco 1509 can be used on aluminum car radiators, A/C evaporators and condensers, and other similar metals.

This is a low-temperature self-fluxing solder alloy for aluminum window frames, zinc-based carburetors and outboard-motor housings. It comes in 1/8" diameter rods and melts at 700°F (371°C).

Silver Brazing w/Gas Welder

All State No. 101 and 101FC Trucote Braze—These are general-purpose silver-brazing alloy rods. They have a high tensile strength of 5,200 psi with a working temperature of 1,145°F (618°C). Either rod may be used to join almost any metal with a melting temperature above 1,150°F (621°C). Number 101 or 101FC Trucote braze rod is especially good for soldering copper, brass, steel, stainless steel or aluminum. These rods may be used with a separate flux or can be bought with a blue-colored flux coating. They come in diameters of 1/16", 3/32" and 1/8".

Welco 200 Braze—This is a high-strength, 56% silver alloy for ferrous and non-ferrous metals. Its bonding temperature is 1,155°F (624°C); tensile strength is 85,000 psi! By contrast, the tensile strength of chrome-moly steel is only 70,000 psi! This silver brazing rod can be used for race car suspensions and airplane wing ribs. It requires liquid flux.

Welco 200 Flux—Ideal for brazing, it comes in a 12-ounce plastic jar in paste form. It can be removed with warm water after brazing.

Brass, Bronze & Aluminum Brazing w/Gas Welder

All State Nickel/Bronze Braze Rod—This rod has a high tensile strength of 85,000 psi. It melts at 1,200–1,750°F (649–954°C), so it's a little harder to work with than lower-temperature brazing rods. Regardless, braze rod does a good job when used properly, for assembling bicycle frames and ornamental railings. It should not be used on chrome-moly steel because it penetrates the grain of the base metal and cracks it.

All State No. 41FC Braze—This is a high-quality rod with a tensile strength of 60,000 psi. It is flux-coated, eliminating the inconvenience of using separate flu. No. 41FC brazing rod is widely used to make bicycle frames, race car frames, and for repairing auto bodies. If exposed to the atmosphere over a period of time, the flux coating will flake off and the rod becomes unusable. Therefore, I buy flux-covered rods for jobs I expect to complete within a few days.

All State No. 31 Aluminum Braze—This is primarily meant for use with thin sheet aluminum such as fuel tanks, oil tanks and truck, race car and trailer bodies. It can also be used to repair aluminum irrigation pipe. No. 31 braze rod should not be used on 2024- or 7075-series aluminum. It comes in 18" lengths and 1/16", 3/32" and 1/8" diameters. It has a high melting temperature of 1,075° F (579°C). A good flux such as All State No. 31 should be used.

All State No. 33 Aluminum Braze—This is for brazing aluminum castings. This filler will repair cracked or broken castings and will fill holes or build up areas that have been worn away or broken off. Use All State No. 31 flux with this rod.

Welco 10 Aluminum Braze—This rod is a 30,000-psi tensile strength alloy used on aluminum sheet. Use Welco No. 10 flux.

Anti-Borax Flux—This flux comes in 1-lb cans and is used for brass brazing of brass, bronze, steel and cast iron. It can be used in paste form by mixing with water. Or, it can be used in its dry powder form by simply dipping the brazing rod into the can after heating the rod with the torch. I like to use the heat-the-rod-and-dip method for best results.

Gas Welding Aluminum or Steel

Welco 120 Welding Rod—This welding rod is a high-quality alloy for fabrication and repair of most weldable aluminum alloys. It is easy to control, and the bead solidifies rapidly, producing a nice-looking weld. Use oxyhydrogen if you plan to do much aluminum welding. Use Welco No. 10 flux when welding with this rod.

No. 1100 Aluminum Welding Rod—No. 1100 is the most common rod to use for all-purpose gas welding with a separate flux. Use Oxweld Aluminum flux No. 725Foo. This flux comes in 1/4-lb jars, and the label states for all aluminum welding.

All State Sealcore—This s a unique tubular aluminum welding rod with the flux contained inside its hollow core. This welding rod is claimed to be the most versatile of all aluminum welding rod for torch welding. It comes in 1/8" and 3/16" diameter rods, 20" and 32" long. It is supposed to be good for field work such as repair of irrigation pipe and farm equipment.

Coat Hangers—In the past, some welders have used coat hangers to oxyacetylene-weld car fenders or anything else they could think of. Don't do it! Paint on coat-hanger wire contaminates the weld. And the alloy in the wire is unknown. Usually, coat hangers are so brittle that they break when you try to straighten them. I suspect they are made of the cheapest steel available. You wouldn't want a weld to crack because of poor-quality filler material.

Welco W-1060 Mild-Steel Rod—This gas-welding rod is copper-coated, but it is a good choice for gas-welding mild steel. It is available in 1/16", 5/64" and 1/8" diameters, 36" long. My last invoice from an Airco dealer called this #7 Steel. It works well for automobile exhaust systems and, if used carefully, OK for gas-welding non-structural 4130 or 4340 aviation steel. But the copper coating contaminates the weld if the rod is not used carefully. See the above photo of a scanning electron micrograph of the copper coating.

Airco or Oxweld Bare Mild-Steel Rod—I use this rod for torch welding, gas welding or oxyacetylene welding 4130 steel. And it's what the Experimental Aircraft Association recommends in their welding schools for welding airplane parts. It comes in 1/16", 5/64" and 1/8" diameter rods, 36" long.

Arc Welding

The drawing on this page indicates how to identify arc-welding rod. Welding-rod diameter refers to the wire size, not the diameter of the coating. Wire diameter is measured easily at the holder-end of the rod.

E-6011—This is the easiest to use all-purpose welding rod for arc welding mild steel with a 220-volt AC buzz-box arc welder in all positions. This is a good welding rod when the work is dirty or oily, and you don't have time to make the job pretty. The E-6011 rod produces a considerable amount of unsightly spatter—small globules of molten metal that stick to the base metal—in the area of the weld. I recommend E-6011 arc welding rod for making repairs on farm equipment.

To weld with E-6011 rod, hold a 1/8" or shorter arc. Move at a steady pace that is just fast enough to stay ahead of the molten slag. For welding overhead or vertical-up, reduce the current setting by one notch on the buzz-box. E-6011 comes in 1/16", 5/64", 1/8" and larger sizes.

E-6013—This is an excellent general-purpose welding rod for use with 220-volt AC buzz-box welders when both easy operation and outstanding weld appearance are important. This rod can be used in all positions. The work must be cleaner than when welding with E-6011 rod. I recommend

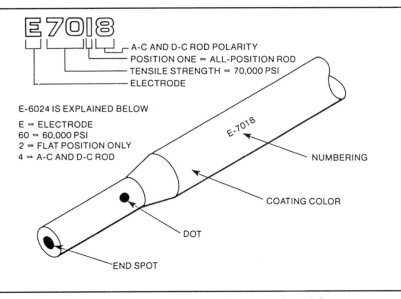

E7018

A-C AND D-C ROD POLARITY
POSITION ONE = ALL-POSITION ROD
TENSILE STRENGTH = 70,000 PSI
ELECTRODE

E-6024 IS EXPLAINED BELOW

E = ELECTRODE
60 = 60,000 PSI
2 = FLAT POSITION ONLY
4 = A-C AND D-C ROD

E-7018

NUMBERING

COATING COLOR

DOT

END SPOT

this 60,000 psi tensile-strength rod for projects such as building trailers.

To weld mild steel with E-6013 welding rod, drag the tip of the rod lightly against the work. Do not hold a gap as you would with E-6011 rod. Some companies call this a contact rod because you always keep it in contact with the base metal. Move steadily and just fast enough to stay ahead of the molten puddle. When welding sheet metal with E-6013, weld downhill. E-6013 welding rod comes in 1/16", 5/64", 1/8" and larger diameters.

E-7014—This welding rod is also designed to be used with 220-volt AC buzz-box welders. Compared to E-6011 and E-6013 rods, E-7014 is a slightly higher-strength welding rod with good appearance characteristics. The first two numbers in the identification indicate 70,000 psi tensile strength. This rod is commonly used for sheet-metal welding.

You should lightly drag the tip of the rod on the base metal when laying a bead. Therefore, you don't have to maintain a gap with the arc. This is a good welding rod for beginners, but so is E-6011. E-7014 comes in 3/32", 1/8" and larger diameters.

E-6010—Sometimes called 5P by oil-field pipe welders, it is an all-purpose, deep-penetrating welding rod for use with 220/440-volt DC welders. It works similar to all purpose AC/DC E-6011 rod, but leaves a much smoother weld and produces almost no spatter. It will work on dirty, oily or rusty pipe and other steel. E-6010 welding rod should not be used with an AC welding machine. Set the machine for DC, positive or reverse polarity when welding with this rod. It comes in 1/8" diameter and larger sizes.

E-7018 LO-HI—Sometimes called low hydrogen, this rod should be used with arc welders.

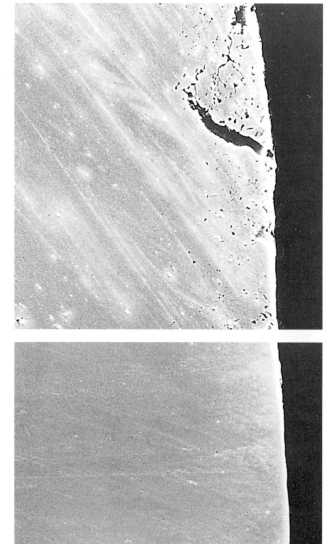

Commercial-grade TIG wire for 4130 steel magnified 1300 times is full of voids and surface imperfections. Compare to next micrograph that shows metallurgically controlled wire; obviously a better product. Courtesy United States Welding Corporation.

Vacuum melt, metallurgically pure, roller-formed (not die-formed) wire, magnified 1300 times, is much smoother than commercial grade 4130 in previous photo. Courtesy United States Welding Corporation. www.uswelding.com

This welding rod was developed originally for 70,000 psi tensile-strength, X-ray quality welds in the nuclear power industry. Regardless, the cost of E-7018 rod is similar to other welding rod, so it is one of the most commonly used welding rods. It produces high-quality, good-looking welds suitable for pipe welding that must be certified. E-7018 rod also works well for trailer frames, race car frames and mild steel.

The work must be thoroughly cleaned and prepared. If you are sloppy, it is easy to bury slag pockets with E-7018 rod. If the metal is clean and properly prepared, the slag will actually peel off as the weld cools. This rod is highly susceptible to moisture damage, so it must be kept dry and clean at all times. Once you get the hang of using E-7108 rod, you'll like it.

EST & Lincoln Ferroweld—Excellent steel welding rods for making high-strength welds in cast iron when no machining is required afterward. These welding rods are used with 220-volt AC welders or 220-volt DC welders set to reverse—positive—polarity.

Maintain a short arc as you would with an E-6011 rod—don't touch the work with the rod tip. It's best to preheat the entire part to 400°F (204°C) prior to welding to minimize stresses. Cast iron should be welded in short, 1"-long beads and allowed to cool in between. This reduces heat buildup and the likelihood of cracking.

E-308-16 Stainless Steel Rod—This rod may be used with a 220-volt welder for certain stainless steel welds where weld appearance is not critical, such as pipe welding. It leaves a finish similar to E-6011 rod—a lot of spatter. Use AC current or DC current set to positive polarity.

Other specialty welding rods designed for use in electric arc-welding machines can be found at your local welding shop:
•Aluminum
•Bronze
•Hard surfacing—a very hard coating of steel applied over mild steel surfaces where abrasive wear occurs, such as bulldozer blades, plow blade and steam-shovel scoops.

TIG Welding

4043—This is an uncoated aluminum welding rod for TIG welding aluminum. I use this welding-rod alloy for most TIG welding jobs on aluminum. It is available in several diameters: 0.020", 0.040", 1/16", 3/32", 1/8" and 5/32". You can weld the following aluminum alloy with 4043 welding rod: 1100, 5052, 6061, and 356 (casting).

Welco W-1200, AWS A5.2-69, Class RG60—This is the most common TIG welding rod for welding steel. If the box of welding rod has a heat number—a reference number from the manufacturer's quality-control department—the particular batch of rod has been checked for quality. If a question comes up about the rod for one reason or another, the manufacturer can trace its history through this number. A heat number usually means the welding rod is for low-carbon, high-strength steel such as 4130.

Don't let anybody sell you copper-coated rod for TIG welding. The copper coating can cause blowholes in a weld and generates fumes that can be hazardous to your lungs.

Vacuum Melt 4130—When you are welding 4130 steel parts that will later be heated-treated for additional strength, the very best rod to use is

United States Welding Corporation 4130-6457V. This rod is certified to content, source of elements, and is the highest-grade welding rod available.

Other welding rods available from United States Welding Corporation are D6AC, vacuum melted, controlled chemistry, primarily used with gas tungsten arc process for critical aircraft structural components and rocket motor cases.

These welding wires/rods are rolled to size, not drawn through dies, and the rolling process produces very smooth wire with no imbedded impurities. Drawing wire through dies usually imbeds lubricants from the dies, which causes defects in the weld. See the 1600 power micrograph pictures for comparisons.

308 Stainless Steel Rod—This rod is what's needed for TIG welding most stainless steel alloys. There are more than 80 different stainless steel alloys; 90% of them are welded with 308 rod! Contact your dealer for special cases, or if you have any questions about which rod to use.

Stainless rod is non-magnetic, of course, and is identified with a small white tag taped to each rod with the identification number on it. This 308 stainless steel rod is used for welding such things as race car and airplane exhaust systems, kitchen cookware and missile parts. Number 316 stainless steel rod is used for TIG welding where salt water corrosion would be a problem, as in ocean-going yachts and power boats. It is also a good rod for welding race car exhaust systems.

SFA 5.16 ERTI-2 Titanium Rod—This rod is for TIG welding one titanium alloy. The rod must exactly match the alloy. Another titanium alloy, Ti-6A1-4V—commonly referred to as 6-4, is the one most commonly used because it's considered the "4130" of titanium. To weld 6-4 titanium, a 6-4 titanium rod must be used.

Titanium rods come in the same sizes as stainless rods. They are also identified with paper tags and usually must be special-ordered from most welding supply shops.

Magnesium Rod—This rod is not carried by most welding supply shops. Because of its limited use, it's a special-order item. Use the same rod as the magnesium alloy being welded, such as AZ92A, AZ101Z or AZ61A rod. The best rod for TIG welding wrought magnesium is AZ61A rod.

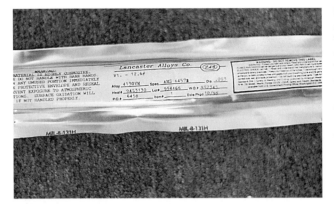

Five- and ten-lb. quantities of vacuum melt, MC grade bare welding rod comes packed in hermetically sealed plastic bags like this 4130 VM rod from Lancaster Alloys Co. See www.lawires.com.

MIG Welding

Electrodes used for MIG welding are smaller than those used in other types of welding. This is because of the high current and speed at which the filler metal is introduced into the weld puddle. For instance, wire sizes start as low as 0.020" diameter and increase in steps of about 0.005". Average size is 0.045" with a normal maximum of 0.090". A maximum diameter of 0.125" has been used in heavy industrial applications. Note: A smaller electrode will achieve more penetration at the same amperage.

E4043, E4146 & #5183 Aluminum Wire Alloys—These are available for MIG welding aluminum. E4043 is most common for shop-welding projects, and for building aluminum trailers for heavy-duty, freight-hauling trucks.

E70S-1, -2, -3, -4, -5 & -6 Steel Alloy Wire—This wire is for welding mild steel. Prefix E indicates electrode. The second digit (7) refers to tensile strength in 10,000 psi, or 70,000 psi. The third digit (0) refers to position—horizontal in this instance. S means solid electrode, versus hollow core. A T instead of an S indicates hollow-core electrode. Dash numbers indicate the chemical composition of the wire as specified by the American Welding Society.

All numbers indicate varying percentages of carbon (C) and silicon (Si) except for -2. The E70S-2 wire also contains titanium (Ti), zirconium (Zr) and aluminum (Al). The higher the dash number, the higher the silicon content.

Chapter 5
WELDING SAFETY

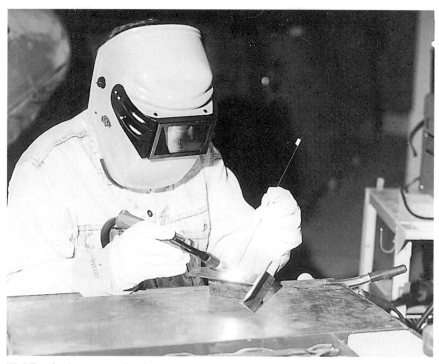

Welding is dangerous business! Make sure you take all safety precautions. This includes proper protective clothing, and the appropriate welding headgear and eye protection required by the type of welding you are doing. Photo: Gayle Finch.

Welding is more hazardous than most other shop processes. The dangers of fire and explosion, burned hands or metal in the eyes are always present. Additionally, high-pressure oxygen, acetylene, argon, CO_2 or helium tanks present a potential work hazard. Hot metal is always a danger if it comes in contact with anything flammable or your skin.

FLAMMABLE CONTAINERS AND WELDING

I have two true stories, among many that I've heard over the years, that I'd like to mention to drive home my point.

The first involved a welder who was assigned the job of cutting some 55-gallon oil drums in half to be used as cattle troughs. He had already cut two, and on the third, he was straddling it like a horse when it exploded. The force was so great that it blew the welder nearly 50 feet straight up through the corrugated tin roof of the building and outside. He didn't make it.

One of my former auto-shop students once told me about an explosion in my own neighborhood. A local oil-trucking company had hired two neighborhood welders to weld taillight brackets on the rear bumper of some new trucks. The trucks were really fancy, with fiberglass cabs, chrome-plated dual exhaust stacks, and a large stainless-steel tank on the back for hauling crude oil. The welding had been completed on one truck. The second truck was backed into the driveway so the welder could get the welding cables to the rear bumper. The truck was so big that it occupied the entire driveway.

The auto-shop student had been watching, but went home for lunch. He lived in a house one block from where the welding was being done. He heard a tremendous explosion and ran outside just in time to see a huge stainless-steel tank fall on top of the house behind his. At first he thought a missile from nearby Vandenberg AFB had exploded and was falling on the neighborhood. Then he realized the debris had come from where the trucks were being welded, so he ran down the street to find the new truck had been totally demolished! One of the welders had been blown across the street, bruised but alive. The other fellow was found inside the kitchen of the house, also bruised but alive.

The truck didn't fare as well. The oil tank, of course, had been blown off the frame, straight up 100 feet or more and had parked itself on the house roof. The new truck was totaled! The two men welding on the truck miraculously escaped with only minor injuries. The explosion was caused by a welding spark igniting vapors escaping from the crude-oil tank.

The lesson to be learned here is that containers of flammable products should *never* be welded on or cut with a torch. It is safest to refuse to weld on or near any such tank or container, even if it has been completely drained and sitting empty for a long time. The vapors and flammable materials can permeate the metal, and even though it smells clean, the vapors may still be present. Even vapors from non-flammable liquids can be explosive under certain conditions. And for sure, vapors from flammable liquids are explosive.

As a welder, people will ask you to weld many different types of containers. You may even want to weld something similar yourself, but the safe thing is not to do it.

If you absolutely must have a tank welded, such as a classic gas tank for a restoration that simply can't be replaced, take it to a professional welding shop that specializes in welding gas, oil and similar flammable tanks. They'll know what preparation and welding procedures are needed to prevent accidents such as those just described. Typically, a professional welding shop will "boil out" the tank in radiator cleaning acid to remove any oil or gas residue, then purge the tank with an inert gas such as argon while welding it. Clearly, these are techniques above and beyond the skills of even talented do-it-yourself welders.

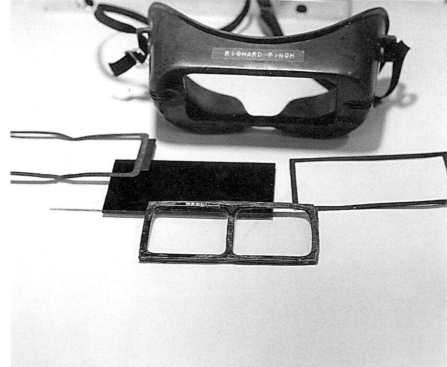

Special corrective lenses (lower left) can be used in place of corrective eyeglasses. Such lenses are available at welding supply stores. This lens will fit either gas welding goggles or an arc welding helmet.

ARC WELDING SAFETY
Radiation Burn

The primary safety hazard in arc welding is from ultraviolet-light (radiation) burns to the eyes and, to a lesser degree, the skin. This hazard applies to TIG, plasma-arc and wire-feed welding as well.

The burns received from electric-arc welding are similar to sunburn, except usually deeper into the skin or eyes. One reason for the deeper and potentially more severe burns may be the ultraviolet light source—it's much closer to the body than the sun is. Therefore, arc-welding radiation is more intense. And there's less atmospheric dust to filter the rays of the arc welder.

Clothing

The solution for preventing burns is simply to shield the skin and eyes from ultraviolet light. You should wear a long-sleeve, close-weave shirt and trousers of material least likely to ignite from sparks. Do not use synthetic materials, such as nylon or polyester, because they melt easily. Regular work clothes are acceptable. Wool or heavy denim, such as blue jeans, works just fine for shielding radiation. Many professional welders use leather aprons, jackets or pants for burn protection. And don't forget welding gloves. Your hands are the closest to the heat and light source. Consequently, they are most vulnerable to burns from radiation,

sparks and hot metal.

For maximum burn protection from your long-sleeve shirt, tuck the sleeves inside the top of the gloves and button the collar at the top. Many beginning welders neglect to button up their shirt collar and end up with a nasty ultraviolet burn on their neck. Wearing a leather flap on the helmet to shield the neck and throat is the best way to avoid neck and throat burns.

Gloves—Protect your hands and wrists from burns with leather gloves. Although leather doesn't burn easily, it will char and shrink if it contacts hot metal or a flame. Choose the most flexible gloves possible. For light TIG welding, I prefer deerskin gloves. For heavy arc welding on structural steel, heavy cowhide or horsehide gloves give you the protection needed.

Shoes—If you weld bridges professionally, you should wear leather, high-top, safety boots. Regardless of whether you weld for a hobby or professionally, never wear low-cut shoes, particularly the slip-on type such as loafers. Sparks and molten metal can slip down inside your shoes. Regardless of your pain threshold, you'll drop everything so you can get your shoe off and the hot metal out while you are dancing the jig. And while you are dancing the jig, the dropped welding torch could start a fire or explosion! Never wear nylon jogging shoes. If exposed to high heat, they could melt to your foot! For hobby welding, an old pair of leather lace-up shoes will be OK.

RECOMMENDED WELDING LENSES

Application	Base Metal Thickness (inches)	Shade No.
Arc welding with 1/16, 3/32, 1/8, 5/32-in. electrodes	1/8" to 1/4"	10
3/16, 7/32, 1/4-in. electrodes	1/4" to 1"	12
5/13/8-in. electrodes	Over 1"√	14
TIG welding with non-ferrous 1/6, 3/32, 1/8, 5/32-in. electrodes*	Up to 1/4"	11
TIG welding with ferrous 1/16, 3/32, 1/8, 5/32-in. electrodes*	Up to 1/4"	12
Soldering	All	2
Brazing	Up to 1/4"	3 or 4
Light cutting	Up to 1/4"	3 or 4
Medium cutting	1" to 6"	4 or 5
Heavy cutting	Over 6"	5 or 6
Gas welding (light)	Up to 1/8"	3 or 4
Gas welding (med.)	1/8" to 1/2"	3 or 4
Gas welding (heavy)	Over 1/2"	6 or 8

For MIG welding, decrease shade no. by one from the TIG suggestion. Make sure your eyes are protected with the right lens when welding.

Note: You should wear eyeglasses that are small enough to easily fit under your gas-welding lenses, and it is easier to weld if you are wearing tinted safety glasses that will fit over your eyewear eyeglasses. Look for shade 3 safety glasses to make gas welding easier for you.

Full face shields are preferred by some welders, but the face shields are a nuisance if someone else is also using them. They must be kept clean to make welding easier to see.

Arc-Welding Lenses

I prefer the rectangular 2 x 4 1/4-in. lens. It's available in various shades and tints. The standard tint is green #10. A darker tint for welding in bright sunlight is green #11 or #12. A #9 lens is OK for low-hydrogen welding or TIG-welding steel. TIG or wire-feed welding aluminum require a darker-tint lens, such as #10, #11, or even #12. You should pick the darkest lens that still allows you to see the weld puddle.

Don't use gold- and silver-plated lenses. Although they are pretty, one tiny scratch in the gold or silver plating could admit enough ultraviolet light to burn an eye.

A tinted lens should always be protected from weld spatter, scratches and breakage. Cover it with a clear, disposable plastic or glass lens protector. Change the lens protector as often as necessary to ensure distortion-free sight. Likewise, clean the lens as often as you would a pair of eyeglasses.

Helmets

Nowadays, electronic helmets are the hot ticket. Every welder who doesn't own one, wants one. They really do work, but there are drawbacks. If you strike a new arc 50 or so times in a day, your eyes will itch at the end of the day if you are using an electronic helmet. That is because the fraction of a second that it takes the lens to darken from #2 to #10 or #12 allows ultraviolet light to strike your eyes. The effect is cumulative. It adds up by the end of the day. The time lag for darkening is noticeable. Another negative is the cost of electronic helmets, which run about 10 times or more the price of a standard arc-welding helmet.

Another special arc welding helmet is a "chin-up" model that allows the welder to lift the lens only, just enough to see the weld area immediately before striking the arc. To close the lens, the welder simply moves his chin about 1/2". For me, this is easier than the other methods of protecting my face, neck and eyes from welding flash burns.

The good thing about electronic helmets is that you can see where the arc will strike immediately before you weld. As with many things in welding, "try before you buy."

There are about 100 different safety-approved welding helmets, or hoods. I have two different helmets for myself.

Get the lightest helmet that covers your face completely. Whether you are welding bridges or ships for a living, or welding a lawn mower or child's bicycle, the lightest helmet will distract you the least. You need all the concentration you can

Clockwise from top left: (1) HTP chin-up arc welding helmet is inexpensive and makes welding easy; just move your chin to open and close the lens. (2) Fiber helmet retrofitted with electronic lens. (3) Lightweight, large lens Huntsman helmet with #10 lens. (4) Daytona MIG/D.Q.F. helmet with adjustable electronic lens. Pick one that suits your budget and that works for you.

eyes. First of all, you should provide a suitable screen around the weld area so direct flash can't be seen by anybody not wearing correct eye protection. For a temporary screen, place a sheet of plywood or corrugated tin so it shields those nearby.

If you can see arc-weld flash from any distance, it can burn your eyes. The closer you are to the flash, the more severe the burn will be. The same rule applies to pets and other animals. Within reason, keep them from looking at the welding flash by using shielding. If anyone is in the welding area, always say "watch your eyes" before you strike the arc. This gives the person a chance to turn the other way or close their eyes. If you weld regularly in a certain area, permanent or portable screens should be used. These can be built or purchased from a weld-supply outlet.

GAS-WELDING SAFETY

Oxyacetylene welding presents its own set of safety considerations. Everything I said about shoes, gloves and clothing for arc welding also applies to gas welding.

muster to do a good welding job, so you certainly don't want a heavy helmet pulling down on your neck and head. Regardless of weight, though, you need maximum protection.

Safety Glasses

If you can't afford safety glasses, you can't afford to weld. Don't think safety glasses are too much bother—they don't compare to the bother of impaired eyesight. Your eyes will not tolerate hot metal in them. I wear my safety glasses when grinding, chipping, filing or anytime there's a chance of metal flying around.

Safety for Bystanders

When arc-welding, you have the obligation to protect bystanders from ultraviolet burns to their

Face and Eye Protection

Goggles for gas welding or cutting are different than arc-welding helmets because they don't have to shield the welder from hostile light. Oxyacetylene welding doesn't produce ultraviolet light, so it can't cause flash burns. The only burn you can get from oxyacetylene is from the heat of the flame, and from sparks from the oxidized metal being welded or cut.

The best goggles for oxyacetylene welding are fiberglass and have a 2 x 4 1/4" lens. You can wear most conventional vision-correcting eyeglasses under them. In addition, they are easier to put on and adjust than the separate-eyepiece goggles.

Bottled Welding Gases

Welding gases come in high-strength, high-

BOOM!

Although acetylene is bottled at comparatively low pressure, the acetylene itself is highly explosive. Like leaking gas from a natural gas pipeline, nothing happens if acetylene leaks into a confined space unless there is a spark or flame to ignite it. Then, it can explode just like a stick of dynamite. Never carry flammable gas such as acetylene or hydrogen gas in the closed trunk of a car. If vapors escape, a major explosion could result. Instead, haul acetylene gas cylinders in an open truck or trailer.

Don't let these safety tips worry you unnecessarily. Every time you drive a car, you are carrying around enough gasoline to do about the same damage as the contents of any welding gas cylinder. Therefore, if you observe proper safety precautions, you should have no trouble.

pressure cylinders. When full, they contain the following pressures at room temperature:
Oxygen: 2200 psi
Argon: 2200 psi
Hydrogen: 2200 psi
Acetylene: 375 psi

If a pressurized cylinder of oxygen or argon welding gas falls and breaks off a valve, the resulting release of high-pressure gas can turn it into a rocket. I know of one high-pressure cylinder with a broken-off valve that went completely through four automobiles and the brick wall of a welding shop. As with fire, bottled welding gas is useful when contained and controlled. But also like fire, pressurized gas can be destructive. Always chain welding-gas cylinders in an upright position to prevent them from falling. An acetylene cylinder must be kept upright to keep acetone, a stabilizing gas, at the bottom of the cylinder. If the heavier acetone gets into the cylinder regulator, it will damage its rubber parts.

GENERAL SAFETY TIPS
Work Area

The most common welding accident is burned hands and arms. Keep first-aid equipment for burns in the work area. And, as I already discussed, eye injuries can occur if you get careless. So keep handy a phone number for emergency medical treatment for unexpected injuries, particularly eye injuries. A medical doctor is the only one who can properly treat any eye injuries. Don't try to treat an eye injury yourself. Get to a doctor immediately.

The ideal welding shop should have bare concrete floors, and bare metal walls and ceilings to reduce the possibility of fire. Although you probably don't have such a building, the important thing is to keep flammables, and rags, wood scraps, piles of paper and other combustibles out of the welding area. The same goes for wood floors; never weld or cut over

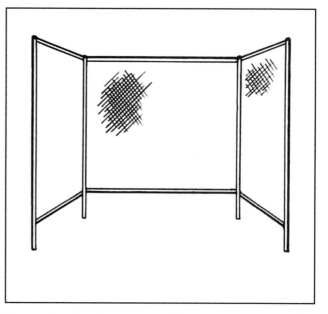

Be sure to do all your arc welding and plasma cutting behind a screen to protect bystanders and pets from eye damage from ultraviolet radiation. You can buy screens at your local welding supply, or you can make a temporary screen from three pieces of 1/8" to 1/4" plywood, hinged together at the edges.

one.

If you must weld in an enclosed garage, make every effort to eliminate anything that could trap a spark. Sparks can smolder for hours and then burst into flames. So, regardless of where you're welding, be sure to have a fire extinguisher nearby. Also, keep a 5-gallon bucket of water handy to cool off hot metal and quickly douse small fires.

Never use a cutting torch inside your workshop. Take whatever you're going to cut outside, away from flammables. Also be aware that welding sparks can ignite gasoline fumes in a confined area.

Grinding Sparks

When grinding with a portable grinder or bench

Inexpensive drug- store-purchased reading glasses will do in many cases to prevent you from having to wear your expensive prescription glasses in the welding shop.

SOLDERING SAFETY

You may have heard stories about the dangers of lead poisoning from soldering. Here's how lead-poisoning dangers can be avoided:

• Some fluxes are toxic.

•Smoking, eating or drinking while handling lead solder can eventually cause lead poisoning. Be sure to wash your hands and face with soap and water after handling lead solder.

•Do not breathe fumes from overheated lead solder. If you keep the temperature between 250°F (121°C) and 900°F (482°C), there is almost no danger of lead poisoning. But lead heated to over 1,000°F (538°C) will give off fumes that are hazardous to your lungs and body.

•The harmful effects of absorbing lead into the human body are largely cumulative. If you experience colic, muscular cramps or constipation when using lead, stop immediately and contact a physician.

grinder, the resulting high-velocity sparks are tiny pieces of metal. These tiny projectiles are in an oxidized state and at the same temperature as metal being cut with a gas torch. Therefore, be as careful of where grinding sparks fall as you would if you were using a cutting torch. There are now full face plastic head shields available that you should wear when grinding to provide maximum protection.

Eyeglasses

An often-overlooked cause of less-than-perfect welding is not being able to see the weld puddle clearly. Frequently, people who don't have good vision attempt to weld without eyeglasses. If you want to do a good welding job, but need glasses, have your eyes checked and corrected first. Or, wear your glasses.

If you don't want to subject your expensive bifocals to welding damage, there are other solutions. The least-expensive solution is to buy a pair of reading glasses at a drugstore or supermarket. These come in standard magnification increments of +1.25, +1.50 and so on. Welding supply shops also sell eyesight-correcting lenses that fit the 2 x 4 1/4" opening in helmets or goggles. These are available in +1.25, +1.50

Keep your shop tidy. Don't let your welding cables lay on the floor to get stepped on and shorted out. Make hooks to hang the cables on and build carts to sit the welding equipment on.

and higher increments, too. You might want to have a special pair of "computer glasses" made to your prescription if you have astigmatism, and tend to see two of everything through your bi- or tri-focals. This should solve the "seeing two of everything" problem for you.

Comfort

I suppose most people think a welder at work is not comfortable. That is somewhat true, but the more comfortable you are while welding, the better your

For all kinds of bare wire welding, gas, MIG and TIG, this pair of pliers are the handiest you will find. They can cut wire cleanly and they can clean out the nozzle and take apart the TIG collets and chucks. Ask for "Yelper YS-50 MIG Plier."

work will be. I prefer to sit when welding. Why stand and have leg fatigue interfere with concentration?

Just as you would sit down and get situated to write a letter, get yourself comfortable before starting to weld. Think about it: You wouldn't squat down and write a letter on the floor. Likewise, you wouldn't stand up with your arms stretched out to write a letter with the paper against a wall. And you certainly wouldn't lay on your back, writing a letter using the side of the desk! If you tried writing in these awkward positions, your handwriting would not look good. That's why you can't weld as well lying on your back, or squatting on the floor.

There are exceptions. For instance, you can't flip a car on its roof just to weld on a new muffler. However, you can get yourself as comfortable as possible if you have to weld in such a position.

Simple plywood-and-nail and wood block jig for 4130 steel-tube airplane frame. Fit tubing so 1/16-in. welding rod cannot be inserted in weld-seam gap. Remember, if the parts don't fit exactly right, you won't be able to weld them properly, no matter how good your welding skills are.

B elieve it or not, this is the most important chapter in this book! You can be the world's best welder, but if the parts fit poorly, you simply cannot weld them properly. If you have a 1/4" wide gap in a piece of .030" sheet metal, you simply cannot do a nice, strong weld across that gap. If you have a rough, jagged angle iron (steel) frame to weld for a two-wheel trailer, the welds will be weak and could crack when least desirable, like when you are towing a heavy load on the highway.

Fitting is the process of cutting and shaping metal pieces so they fit together without large gaps. Big welding operations usually employ a welder-and-fitter team–two people who fit and complete the welds. If you were to time and apportion the operation, fitting takes about 80% of the time; welding takes the balance. These percentages illustrate how important fitting is to making good welds.

You must fit pieces properly before welding to make sure they stay in place during welding and subsequent cooling. The idea is to avoid large gaps that must be filled with weld bead.

Jigging is assembling a project in a fixture to ensure that the welded assembly conforms to design specifications. Jigs are used in large-scale production to ensure consistent quality. You must plan ahead of doing the actual welding in order to get good welds.

FITTING
Thin-Wall Tubing

When welding thin-wall steel tubing, such as for a race car roll cage or an airplane fuselage, I fit the tubing joints so there are no gaps—the joints are almost watertight before welding! At the very least, your welding project's joints should fit close enough so a welding rod cannot be inserted

Roll bar and race car frames need to have the tubing ends notched as is being done here with a Williams Lowbuck tubing and pipe notcher.

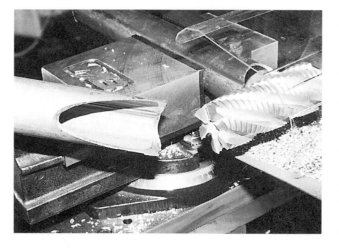

Mittler Bros. Tool Co. makes this tubing notcher that can notch up to a 70-degree cut.

Up to 2-inch diameter tubing can be fish-mouthed in a matter of seconds. This notcher bolts to the front of your workbench and is operated with a 1/2-inch drill motor. Dale Wilch Sales.

between them. See the photo on page 37.

One of the toughest fitting jobs is building a set of tubular engine-exhaust headers. For welding, fit the weld joints as closely as possible. It's not easy. Even though I've had a lot of practice, I sometimes over-trim a curved section of exhaust. If I over-trim a tube, I have to splice in a section or start over with a new piece. Practice is still the best way to get a good fit.

Always cut the pieces a little longer than necessary so you'll have plenty of metal to file and fit into the proper shape. As one of my welding supervisors used to say, "You cut it off twice and the piece is still too short!" This means that if you start with a metal piece that's too small, you'll never be able to fit it properly by trimming. So, be sure to rough-trim a part first to provide extra metal for final fitting. The resulting fit will be much more accurate.

Fishmouth Joints

Recent improvements to long-established fish-mouth fitting methods have greatly simplified this tubing fitting process. In this chapter, I have illustrated several different tools that make

fishmouthing and fitting tubing a lot easier than it was even as recently as 1990. If you have lots of tubing to weld, by all means try one of the tools I have illustrated here.

The fishmouth tool that uses bi-metal hole saws will work very well on all kinds of tubing from 1/2" diameter up to 2 1/2" diameter. Larger sizes could be cut by making a larger tool and using larger diameter hole saws. I have cut 4" diameter tubing by making special setups in a vertical mill. There is one limitation to the hole saw method, and that limitation is the angle of cut. Most hole saw cutters will do 90° cuts down to 45° cuts, but no steeper angles.

The milling cutter shown in this chapter will mill cut angles down to 28°, but the cost of the tool is a lot more than the cost of a hole saw cutter.

If you don't have a lot of tubing fits to make, you can still do the job by hand. I use a half-round mill file to hand-fit most tubing. A small air die-grinder and rotary file will also work in many cases, but this method is very slow compared to the power cutter method.

Angle and Plate Steel—Even when fitting angle and plate steel, you need a tight fit. Any wide gaps will vary weld quality, making the finished joint weaker.

Heavy-Gauge Material—When fitting thick-wall tubing, heavy pipe or steel plate, you must bevel the edges to provide proper penetration. Some certified welding specifications call for the gap to be less than one-third the diameter of the welding rod being used, regardless of base-metal thicknesses. That's a tight fit!

Tools

Tools for fitting range from the ubiquitous hacksaw to a heavy-duty power shear. The hacksaw is obviously a hand tool; the shear a power tool. You'll need some tools from each category to do a quick and accurate job of fitting. You'll also need some tools to complement your cutting tools.

Measuring Tools—These are needed to establish where a cut will be made. One of the handiest measuring tools is a retractable steel tape. A 10' tape will usually do; however, big projects can require a 50' tape. For smaller jobs, you can use a 12" or 18" rigid steel rule. A 6" flexible rule in your shirt pocket can come in handy.

Also, a carpenter's framing square is necessary for laying out those square cuts. One of these durable tools is great for marking parts to be cut or fitted, or setting them up to be welded.

In addition to a framing square, you should also get a machinist's combination square. Use this tool

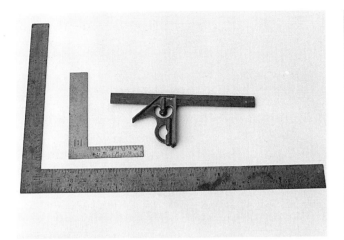

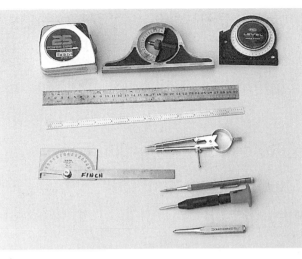

Tools that are necessary for weld parts measuring and fitting. Left Photo: Carpenter's framing square, cabinet square and sliding 45° and 90° scale. Right Photo: 25-ft. steel tape, machinist's angle finder, carpenter's angle finder, 12" steel rule (inches and mm), 12" steel rule (in 100ths), angle protractor, circle scribe, 2 spring punches, steel punch.

for fine layout work and for work involving angles other than 90°.

Marking Tools—These are needed to provide a line for making cuts according to your measurements. When making a cut with a torch, you'll need a mark that won't be obliterated by the flame. These marking tools include a scribe and a center punch or prick punch. A center punch will work, but a prick punch is made specifically for this purpose—making a line of closely spaced prick marks.

Another method for marking metal to be cut with a cutting torch is with a soapstone marker. It marks like chalk, but will not burn at high cutting temperatures. By all means, have several pieces of soapstone marker in your toolbox.

For marking arcs or circles, you'll need a divider—a special scribe that's similar to a compass, but with two sharpened steel points. One place you might want to use a scribe mark and bluing is for flame-cutting flanges for exhaust manifolds. Here, the cut must be precise, and the scribe line is better than soapstone for accuracy.

Remember that you should only use a scribe for marking a cut line. This is particularly true when marking sheet metal because of the tendency of a tear or break to start at a scribe line—the scribe line creates a stress riser.

SOMETHING NEW IN FITTING TUBING!

There is really something new in welding up tubular structures, whether it is an airplane engine mount, a complete fuselage, a race car frame or a special suspension member. A company in Ontario, Canada, has designed a computer-operated machine that can cut to accurate lengths of tubing and then cut accurate trimmed profiles for each tube so that there is no waste except for the cuttings that come off as the 1/8" cutting tool trims the tubes.

Pictured here is a computerized fit-up of thin-wall 4130-N steel tubing that is offered by a company located in Stratford, Ontario Canada www.vr3.ca or you can call them at 519-273-6660. Courtesy Don VanRaay, VR3 Engineering Ltd.

A number of aircraft designs have already been developed into kits that have each tube marked to show where it will fit in the completed structure of the kit. Some of the airplane frames available as a kit are: Bearhawk, One Design, Wittman Tailwind and the Steen Skybolt. The same type of kits could just as easily be done for kit cars and race car frames.

In addition to fitted straight tubes, bent tubes can be done to curved tubes from 2 degrees to 120 degrees. Tubing lengths from 1" to 20' can be profiled and fitted. Because the computerized fitting process is so simple, it is possible to fit one-off designs nearly as inexpensively as doing 100 or more fittings of kits.

As opposed to previous plasma cut fittings, there is no heat hardening in 4130 chrome-moly tubes.

This tubular airplane fuselage was assembled from tubes fitted by VR3 Engineering. You simply submit a 3 view, dimensioned drawing of your tubular structure to them and they cut the tubes to perfect lengths and fit them for you, marked and ready to fit and tack weld. Courtesy Don VanRaay, VR3. www.vr3.ca

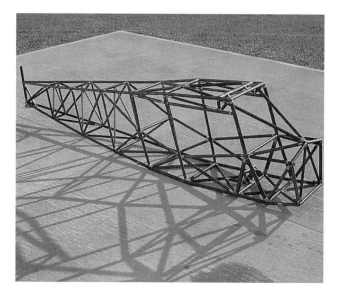

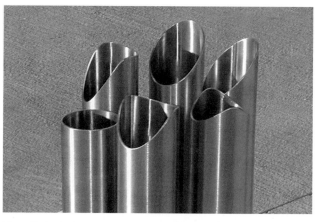

The finished ends of these fitted tubes show how accurate the VR3 company can fit your tubes, all the way from 1/4" diameter to 2 1/2" diameter, which would work just great for building numerous duplicate race car frames to accurate dimensions. Courtesy Don VanRaay, VR3. www.vr3.ca

If your project calls for tubing made from square or rectangular tube such as is shown here, you can simply send a 3 view, dimensioned drawing to VR3 Engineering and you can have ready to weld parts like this. Courtesy Don VanRaay, VR3. www.vr3.ca

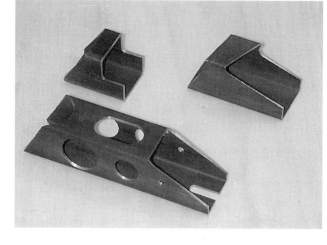

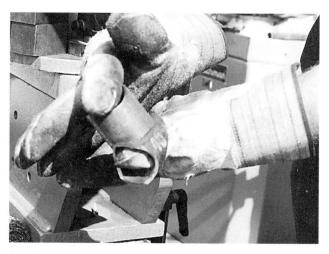

The Scotchman metal shear can notch heavy black pipe for ready-to-weld fit-ups.

This process makes for less stress in the finished weldment. And since all fits are so close, there is no air gap to cause bad welds.

As has been covered here in this chapter, the fitter needs 90% of the time to make a weld complete and the welder only needs the other 10% of the time to finish the weld. Therefore, having a computer guided fitting tool can save about 90% of the time required to build a tubular structure. This can make significant savings in a production manufacturing environment and can save significant time in even one of a kind tubular welding projects. It would pay you to contact VR3 Engineering Co. before you make the first cut of your tubular welding project.

Hand Tools—As mentioned previously, the hacksaw is the first to consider. Just make sure you have plenty of replacement blades. A hacksaw is great for cutting tubing and small structural shapes such as angle iron and channels. Tinner's snips are great for making straight cuts on sheet metal; make curved cuts with aviation snips. Aviation snips can also be used for trimming thin-wall tubing. A plasma cutter can be very useful in trimming thicker parts for welding.

For final fitting and smoothing, you'll need an assortment of files. Start with three basic styles: flat, half-round and round. Get double-cut, coarse-tooth files for removing the most material with each pass. As for the size of the files, bigger projects require bigger files, and vice versa.

Power Tools—Professional welders prefer power tools because time is money, and these tools remove material quickly. Even though the initial tool cost is higher, time is worth more. If you weld much angle steel, a 4" rotary, hand-held grinder is invaluable. Your welds can look totally professional if you dress

Tubing will flatten if you try to bend it without the proper tool. This hand-operated mechanical bender will make correct bends in tubing up to 2" and .139" wall thickness. ProTools photo.

The proper way to make 90° fit-ups in angle steel. This fit-up was made in just seconds with the Scotchman metal shear.

Magnetic angle clamps really come in handy for making fit-ups of steel parts. They don't work on aluminum, copper, brass, stainless and other non-ferrous metals.

For tubing bends that require angle accuracy, this 20-ton upright hydraulic bender will make bends that are repeatable by observing the bubble level at the right on the tubing. ProTools photo.

them with a grinder. A grinder is also great for fitting and dressing after using the cutting torch.

There are many power tools from which to choose. One of the most common is the disc grinder. It's great for smoothing cuts, especially those made by a torch. A disc sander can be fitted with a cup-type stone, which removes metal quickly. For smaller fitting and smoothing jobs, a 2" grinder works particularly well. It's easier to control, thus, a more accurate fitting job can be done.

Die grinders—pneumatic or electric—are useful for making accurate fits. Abrasive stones or carbide cutters can be used, depending on the material being fitted. For instance, a carbide-steel cutter is useful when fitting soft metals, such as aluminum, because of the tendency of an abrasive to load up–clog with metal particles. When coated with paraffin wax, a cutter has less tendency to load up.

For making fast cuts, a saber saw can be used in place of the hacksaw. The powered, oscillating saw rapidly eats through thin-gauge metal, and must be used with care. It takes more effort to follow the cut line than to make the actual cut.

For making cuts on heavy-gauge steel, it's frequently best to use a cutting torch (see page 74). Because flame cutting leaves a relatively rough edge—how rough depends on the skill of the operator—it's always necessary to do the final fitting and smoothing with a grinder.

Safety—When using cutting and trimming tools, protect yourself against flying metal chips. This is especially important when working with power tools, such as high-speed grinders and cutters. Protect your eyes. Clear goggles are OK, but a full-face shield is much better.

Don't forget your arms and legs. Wear a shirt and

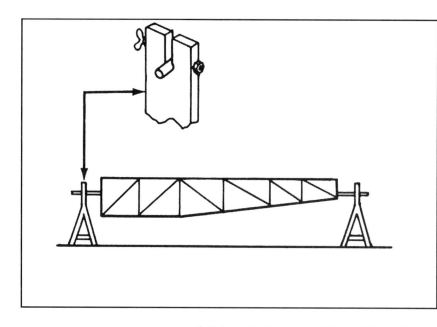

Simple rotating fixture brings the weld seam to the welder. Using one saves a lot of time and makes all of the welds easy to see and to weld.

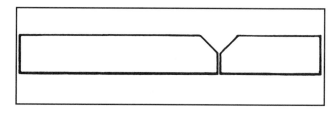

It is not possible to weld thick materials with one pass. You must bevel the edges as shown here, then weld two, three or more passes to obtain 100% weld fusion and strength.

pants with full-length sleeves and legs. Wear gloves when using a disc grinder or torch.

WELDING JIGS

A welding jig is a fixture designed for holding parts in position during welding. The term scares many would-be welders because they envision a very complicated device designed by a welding engineer. This is true only if it was designed to weld hundreds of parts on a production-line basis. Jig-welded usually means that all weldments, or welded assemblies, come out looking exactly the same, with consistent quality. For example, you should use a welding jig if you're building 1,000 airliner seats.

For a onetime project, a jig can be as simple as Vise-Grip pliers or just a piece of angle iron used to prop up a part while it's welded in place. The more sophisticated welding jig can be like that shown above for welding race-car frames and roll cages.

Welding jig for fuselage of steel-tube airplane: I use 5/8"-thick particle board and nail 1 x 1-1/2 x 2-in. pine blocks to it to hold tubing in place. Position blocks about 3 in. from each weld joint to avoid fire hazard.

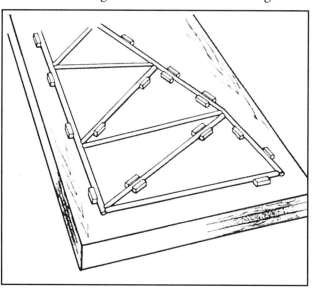

Fortunately, you probably won't have to concern yourself with such a device. The most frequently used jig in my welding shop is a simple three-legged finger. The metal finger takes the place of your finger so you won't get burned while holding a part in place during welding.

Remember, a jig is whatever it takes to hold parts in place until tack welding and finish-welding can be done. Tack welds are a series of very short welds spaced at even intervals. The tack welds hold two pieces of metal together so they can be finish-welded.

Wooden Jigs—A wooden jig is just that—plywood or particle board with small wood blocks nailed to it to hold metal pieces in place. When welding one or two assemblies, such a simple wooden jig is sufficient. See the illustration below. I have welded airplane parts, race-car parts, even factory production parts such as turbocharger wastegates, turbocharger control shafts and seat headrests on wooden jigs.

Permanent Steel Jigs—Factories use heavy steel welding jigs to ensure consistent sizes and fit of parts. You wouldn't want to buy a new exhaust system for your car and discover that the factory had welded the muffler inlet pipe on the wrong side. Factory welding jigs to ensure that welded parts will be interchangeable. Even small race-car factories have welding jigs to assure interchangeability and to improve production rates.

Welding jigs have their drawbacks, though. For example, when welding 4130 steel in a heavy welding jig, the parts sometimes must be stress-relieved to remove internal loads. The jig doesn't allow the parts to twist and conform to stresses from warpage. It holds the weldments in position, regardless of how they want to move. Therefore, internal stresses develop in the welded assembly. These stresses must be relieved, or the part may fail when it's put into use, or service. See the sidebar on page 44.

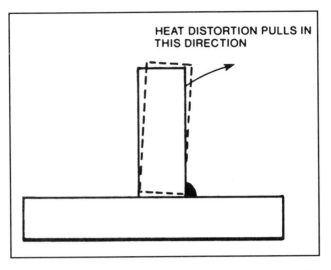

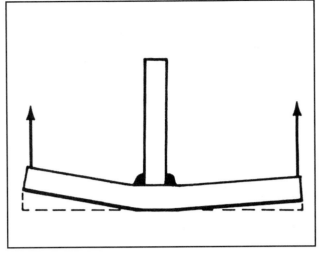

Metal tends to shrink when welded. As weld bead cools, the vertical piece in this drawing is pulled toward the weld. Allow for distortion by tack-welding both sides, then welding on alternate sides as illustrated in stitch-welding, page 64.

Horizontal piece warps as weld bead cools. Such warpage is normal, and can be corrected by heating and straightening after the weld is completed. In this illustration, dull red heat should be applied to the bottom of the horizontal plate to straighten it.

Tips for Welding With Jigs

Avoiding Warpage—Study the sketches in this chapter and you'll learn to cope with warpage in welding. I said cope with because you cannot stop warpage, just limit it.

When building a large tubular structure such as an airplane fuselage, I usually start welding at the front and work toward the back, alternating from side to side. Even better would be to have two people welding symmetrically opposite sides simultaneously, but that is not done easily. So do the next best thing when welding a large structure: Weld one joint, then the opposite one to cancel the effects of warpage from welding the first joint.

Check alignment after each pair of welds. Repeat this welding-and-checking process until the structure is completely welded. It would be a shame to get a frame or fuselage 80% complete and discover that it's 1/2" out of square. A frame, fuselage or large assembly that far out of square is scrap metal.

Tack Weld First—Almost every structure should be tack-welded prior to finish welding. As mentioned, tack welds are a series of small welds between two adjacent pieces. Spaced about 1 1/2" apart, they serve to align the two pieces, hold them together and help prevent warpage. When the final bead is made, the tack welds are remelted and become a part of it.

Only where the designer calls for complete welding of a joint before welding another section should you bypass the tack-welding rule.

Gas Welded Structures—When torch-welding

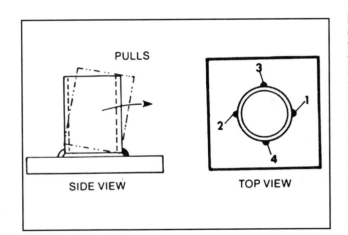

Example of controlling warpage from tack welding: Space tube up from flat plate. Make tack-weld 1. Then square tube to plate; make tack-weld 2. Make tack-weld 3. Then square tube to plate; make tack-weld 4. Finish weld can now be made.

an airplane fuselage or race car frame, I stress-relieve the just-completed weld by slowly pulling the torch away from the work. I do this over a period of about 60 seconds. I never just finish a weld and jerk the torch away. That would surely cause cracking.

Another way to minimize stress cracks in welded assemblies is controlling the air temperature in your workshop. Never weld in a cold or drafty workshop. A weld is more sensitive to cold and drastic changes than the human body! You'll have the best results welding in a room temperature of 90°F (32°C), but 70–80°F (21–27°C) works OK, too. All you'll have to do is become accustomed to working at above-average room temperature.

Don't try to weld in extremely cold weather. The chances of 4130 steel cracking after welding at 40°F (4°C) are 20 times greater than at 80°F (27°C). Mild steel is less prone to cracking from cold-air

STRESS-RELIEVING WELDED ASSEMBLIES

When a complicated, rigidly braced structure such as an airplane engine mount, powerplant high-pressure steam pipe or a race car suspension member is welded, stresses remaining in the metal can cause premature fatigue cracking—caused by many loading and unloading cycles—unless they are relieved. Stress-relieving is accomplished by heating part or all of the structure to about two-thirds of the melting point and then holding the structure at that temperature for two hours or longer to assure that all the residual stresses are relieved. Then the structure is allowed to cool to room temperature in still air.

It ought to be obvious that stress-relieving is not possible except in an oven. However, many long-time welders falsely believe they are stress-relieving structures or parts of structures when they heat it with an oxyacetylene torch and then immediately let the structure cool. They are *annealing*, softening the part, taking away tensile strength, but they are not stress-relieving the part. In most cases, parts heated to blood red with a torch and then allowed to cool immediately are weaker and more prone to cracking than if they were just welded and left to cool. If you still don't believe me, then check out what I have to say about the actual process in Chapter 8, page 73, then decide if you have the tools, equipment and skill to stress-relieve in your own shop.

Or, if you have real need to stress-relieve a part or an assembly that you have welded, by all means, contact a MIL-SPEC, certified heat treater in your locality for advice.

Thankfully, most 4130 steel welded structures are pretty strong and won't crack in spite of how badly they are treated by improper post-heating.

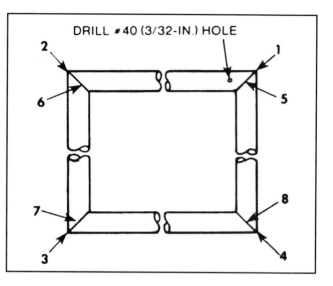

Sequence used to tack-weld tubing in rectangular pattern. Setup is similar to aircraft-fuselage or race car frame bulkhead. When welding closed tubing, vent by drilling hole as shown to prevent weld from blowing out as it's finished.

shock after welding.

TIG Welded Structures—When I am in a hurry to TIG weld something, it is often hard for me to remember to let the post-flow timer do its thing, which is to protect the weld from air until it has cooled to a certain temperature. But that is exactly what the argon post-flow timer is for, to protect the weld just completed. Let it work for you. Don't weld and jerk the torch away.

DESIGNING AND BUILDING WELDING JIGS AND FIXTURES

Much is said about welding out of position. You've probably seen bumper stickers noting the various positions welders are capable of performing in. Although position in welding terms means the position of the weld surface—flat, vertical or overhead—all experts agree that the welder should be in as comfortable a position as possible. So, avoid welding while laying on your back or

standing on your head if there's an easier, more comfortable position.

The way to get in the best welding position—for both you and the electrode—is to make a fixture for rotating the assembly. If you're welding an airplane fuselage, make a fixture such as that shown nearby. Similar fixtures can be fabricated for welding other assemblies. With this fixture, you have better access to all welds. If you're welding a large trailer, hoist it up on its side and turn it over to get to the other side of the weld joints. The "fixture" doesn't have to be exotic. The secret is, make the work accessible to you.

Obviously, it is not possible to make a weld position fixture to flip an ocean liner up on its side for welding, but every effort should be made to position your parts for FLAT position welding of all seams.

Almost anything you use to hold parts in the proper fit-ups for welding would be considered to be a welding fixture. Making a fixture for one-time weld projects is the bottom of the scale, and building massive weld jigs to produce hundreds of identical welded assemblies is at the opposite end of the scale. Most production weld jigs weigh 10 to 20 times more than the parts they produce. In the next several paragraphs I will tell you how to design and build five different kinds of welding jigs.

Plywood, Nails and Wood Blocks Jig

Look at the photo on page 37. You can see that I drew a simple outline on the plywood of the tubing

After 20 hours of cutting, fitting, welding and painting, the Buick V-6 engine mount is bolted to the airplane firewall, ready for the engine.

I use inexpensive electrical conduit—EMT.—to mock-up engine mounts for auto engine-to-airplane installations.

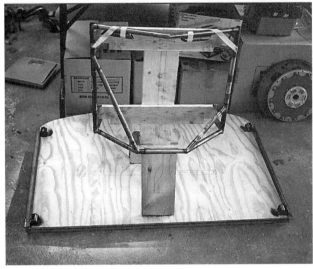

Once the EMT tubing mock-up is completed, the real 4130 steel tubing parts are cut and fitted to the plywood firewall mock-up.

shape I wanted. Next, I drove small finishing nails into the plywood to help position the two pieces of 4130 steel tubing. Because the plywood will catch fire easily, I wet the plywood with water in the tack weld area, and then I only tack-welded the tubing together while it is on the plywood. The plywood often chars if I am gas welding, and it will slightly char if I am TIG or MIG welding. When you are using plywood or particle board for welding jigs, just make sure you thoroughly wet the charred area to prevent the board from catching fire and burning.

Particle Board and Blocks

Many airplane and many race car frames have

been welded together on a wooden welding table like the one shown in the photo on page 37.

Particle board is good for building large welding jigs because it is naturally flatter, smoother and less

I use a lot of corrugated cardboard to make patterns for 1/4" thick steel parts, as shown here. The photo at right shows the finished part.

The 1/4" steel weld assembly in the photo at left is this wheel cradle that makes hangar parking of my airplane much easier to do. The tow bar is mild steel, welded.

Jess Meyers and his associates have built a temporary engine support framework to hold this 4.3-liter Chevrolet V-6 engine in place while they develop the engine mount for the RV-6A experimental airplane. Belted Air Power photo.

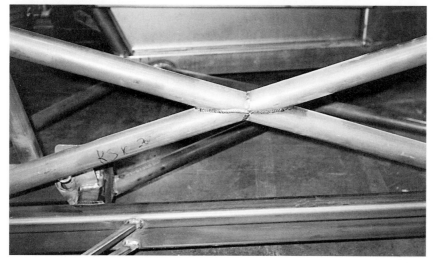

The angle of this tubing fit-up is 55°, too steep for most of the hole saw type notchers, but a notcher cut can be made, then enlarged by hand filing and hand grinding.

Plywood and Bolts Jig

Building a dimensionally accurate tubular framework such as an airplane engine mount requires a welding jig that will maintain the mounting dimensions and bolt pattern of the airplane fuselage. In the pictures shown on pages 45, you can see how I designed and built a plywood-and-bolts welding jig for my airplane project.

First, I drew an exact shape of the airplane firewall on a piece of 3/8" plywood and cut out the plywood firewall shape with a saber saw. Next, I made four exact replicas of the four engine mount brackets that are part of the airplane firewall. Then I bolted the four brackets to the plywood and backed them up with 2 lengths of 1" x 1" angle steel for stiffness. The 3/8" plywood would twist if not braced with the angles.

Then, I fitted and tack-welded the actual engine mount brackets and, with the engine in place, I fitted galvanized EMT tubing to the firewall jig and the engine. Once I was satisfied with the engine mount design, I removed the EMT framework and fitted 4130 tubing to make the actual mount. Total time invested in building the firewall jig, fitting the engine to the firewall, and tack-brazing the EMT mock-up mount was about 25 hours. Building the actual 4130 engine mount took about 12 hours.

Once the mount was completely welded on the plywood-and-bolt jig, it was removed from the jig and fit-checked on the airplane firewall. Three of the four mounting bolt holes slid into place and fit perfectly the first time. The lower right side mounting point was off by about 1/8", probably because my plywood jig was not 100% accurate. Good fitting procedure, combined with TIG welding, produced a strong accurate engine mount.

Surface Plate

A favorite way of making welding jigs for many welding shops is to buy a large, heavy sheet of mild steel. Most shops use 4' x 8' x 1" thick steel plates that weigh 1,325 pounds. A 4' x 8' x 1/2" thick

expensive than similarly sized sheets of plywood. The framework under the particle board should be 1" x 4" lumber, as straight and knot-free as possible. Remember that a crooked jig will produce a crooked part.

The particle board welding jig table should be made slightly larger than the framework you are building. You will need at least one inch extra on all sides so there will be room to nail the 1" x 1" x 2" positioning blocks. The race car frame shown in the picture on page 85 was built on a single 4' x 8' x 3/4" sheet of particle board. Building the welding jig took about 4 hours and fitting and tack welding the tubular race car frame took another 12 hours. This table method sure beats trying to build a race car or airplane frame on a concrete shop floor.

sheet of mild steel weighs 662 pounds.

Most fabrication shops support the steel plate on a framework made of 1" beams and 2" heavy wall tubing, with adjustments on each of 6 or 8 legs so that the table can be adjusted to make it perfectly level. Having a level weld table makes it easy to level and fit parts on top of the table.

Some weld shops allow the welder to tack-weld his assembly to the top of the table. When the assembly is ready to be removed from the table, the tack welds are cut off with a thin abrasive wheel, and then the steel tabletop is dressed smooth with a #80 grit flap wheel.

Other welding shops do not allow tack welding on the table top, so parts must be clamped to the top with C-clamps and magnetic holding devices. You may want to have a dozen or more of these welding aids in your shop for making welding fixtures.

Thick steel plate can also be purchased in larger slabs, up to 8' x 16'. A one-foot-square piece of 1" thick steel plate weighs 42 pounds. It is also possible to buy special cast iron welding tables that are made with up to 50% of the table top open for jigging and clamping.

Butler Race Cars of Goleta, CA, uses this strong steel jig to position parts for welding AC Cobra car replica frames.

Tubular and Angle Jigs

If your plan is to make a large number of welded assemblies, you will certainly want to build one or more permanent weld jigs. A typical application of this weld jig might be for building replica kit car frames like the AC Cobra shown on this page.

The easiest way to build a welding jig of this complexity is to first build a prototype race car frame and attach all the components and brackets that would be included in the finished car. Next, take the car apart and set the frame up at a good, workable height. From this prototype frame, you can develop a welding jig that will allow you to place each part in it for welding.

Often, the jig to build a 200-pound car frame in will weigh 2,000 pounds by itself. Each jig design will be unique to the parts you need to build, but most important of all:

Make sure the welded assembly can be removed from the jig when the welding is done!

We have all heard about building boats in basements, then they won't fit through the exit doorway after they are finished! You will likely need to make the jig a bolt-together fixture so that you can take it apart and put it back together each time you weld up the assembly in it.

For more accurate welding jigs, install locating pins, 1/4" diameter, into each removable connection. You want any and all bolt-together parts on your weld jig to fit exactly the same every time you reassemble them.

HOW TO CONVERT A WOOD-CUTTING BANDSAW TO CUT STEEL & ALUMINUM

Almost any belt-driven bandsaw can be converted to cut metal for fitting and building jigs. There are two secrets to cutting steel on a bandsaw. First, buy a high-quality BI-METAL bandsaw blade with lots of teeth per inch for cutting steel. Second, slow the blade down to a very slow cutting speed, usually 5% of the wood cutting speed.

Correct metal cutting speed for bandsaws is measured in "feet per minute" of blade travel. If your bandsaw has 12" diameter drive wheels, it travels 3.14 x 12" each revolution (12" x 3.14 = 37.68" travel per revolution). 4130 chrome moly steel should be cut at 270-feet-per-minute blade travel. Multiply 270 feet per minute x 12" and you get 3,240 inches per minute. Divide 37.68" bandsaw drive wheel circumference into 3,240 inches, and you get 86 rpm. That is what rpm you need to have your drive wheel turning.

Butler Race Cars' fit-up of rectangular tubing is done by saw-cutting the tubing, de-burring the cut edges. The fit-up is then tack-welded as shown here.

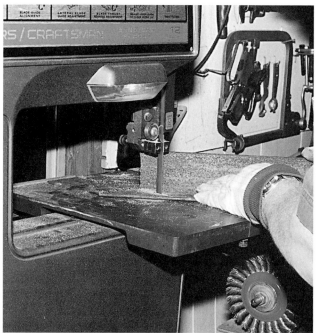

Almost any belt-driven bandsaw can be converted to cut metal. The secret is a quality saw blade and a speed reduction drive that slows the blade down to the right speed for the particular metal. See chart on next page.

Your bandsaw motor probably turns 1,725 rpm. Therefore, you need to gear your motor to the drive wheel down with a 20-to-1 reduction of some sort. I have three different speed ratios on my bandsaw. One is 2-to-1 for sawing wood at 862 rpm. x 37.68" per revolution = 32,480 inches per minute 12 = 2,706 fpm. To effectively saw aluminum plate up to 6" thick, gear your bandsaw to 3.4-to-1 reduction ratio. This gives you a blade speed of 1,593 fpm. To effectively saw mild steel and chrome moly steel up to 2" thick, you need to gear your bandsaws to a 34-to-1 reduction ratio. This ratio gives you a blade speed of 159 fpm. That is somewhat slow for production steel sawing, but your blades will last longer. A 1/2" wide blade works best for steel, a 1/4" wide blade will do for aluminum sawing.

BANDSAW SPEEDS, METAL CUTTING
1725 RPM MOTOR, 12" DIAMETER DRIVE WHEEL

Metal	Blade Speed in fpm	Motor to Drive Wheel Reduction Ratio	Blade Teeth Per Inch	Blade Width
Stainless Steel	Not Recommended	n/a	n/a	n/a
Titanium T1-6A-14V	45 fpm	14 rpm, 125-to-1	18 teeth	1/2"
Bronze	80 fpm	25 rpm, 70-to-1	12 teeth	12"
4130 Chrome Moly	270 fpm	86 rpm, 20-to-1	14 teeth	1/2"
Carbon Steel 1020	330 fpm	105 rpm, 16.4-to-1	14 teeth	1/2"
Aluminum	1,600 fpm	509 rpm, 3.4-to-1	9 teeth	14" to 1/2"
Wood	2,700 fpm	2-to-1	6 teeth	1/4" to 1/2"

For all metals, lubricate the blade with dressing wax stick.

Given: 1725 motor rpm, 12" diameter drive wheel, 37.68" circumference, 12 = 3.14 feet, one revolution, blade travel is 3.14 feet

HOW TO "GEAR" YOUR BANDSAW TO CUT STEEL

The misnomer "gear" should be clarified to explain that belts, pulleys, chains and sprockets are an easy and inexpensive way to slow down the bandsaw blade. You could "gear" down the bandsaw drive wheel by putting a 14-tooth sprocket on the motor and a 280-tooth sprocket on the bandsaw shaft to saw 4130 steel, but you would also need a 1,725 tooth sprocket on the bandsaw shaft to saw titanium, and that many teeth on a sprocket would be very bulky and very expensive. So, you "gear" your bandsaw down in steps.

Employing 3 jack shafts with reduction ratios of approximately 3-to-1 will provide a final drive speed of 65 rpm, more than slow enough to saw chrome moly steel, which requires 86 rpm or less. The first jack shaft should be a V-belt and pulleys to prevent chain and sprocket noise at the speeds that electric motors turn (1,725). A 2" pulley on the motor and a 6" pulley on the first jack shaft will give a 3-to-1 reduction ratio and a speed of 575 rpm.

A 12-tooth #35 pitch chain sprocket on the first jack shaft and a 48-tooth #35 pitch chain sprocket on the second jack shaft, connected to a 24-tooth sprocket on the bandsaw drive wheel shaft, will give 72 output rpm, a final ratio of 24-to-1.

Chapter 7
CLEANING BEFORE WELDING

A strip of open-weave abrasive is used to clean aluminum tubing prior to welding.

Beginning welders think that the heat from welding will burn away any dirt, oil and paint, but that is not so. In fact, dirt, rust, oil, paint, mill scale, oxides and other contaminants will do just that: contaminate the weld, even on farm and ranch equipment.

As a kid on the farm, I believed that welds were expected to break, because they often did if the repairs were done in a hurry, with little or no pre-weld cleaning and preparation. During my years as a welding professional, I learned that good welds were the result of good fitting practices, and proper cleaning of the metal and the filler rod before the welds were done.

CLEANING METHODS

A good rule of thumb is that if it is worth your time to weld something, it is surely worth your time to prepare the weld area before welding to ensure that the weld will be as strong as possible.

Sandpaper Cleaning

For structures that are mostly tubes welded together, sandpaper makes the best cleaning agent. In my shop, I keep a large roll of emery cloth #80-grit to prepare both steel and aluminum tubing for welding. Tear off a length of cloth and use it like shining shoes, usually a clean area about 1" back from the weld area is necessary for preparing the welding. Yes, it takes time to deoxidize aluminum tubing prior to welding, but the weld will be much easier to do with clean metal, and the integrity of the weld will be much better with all the dirt and contaminants removed prior to welding.

If I am gas- or TIG-welding a complete airplane fuselage, I will expect to spend about four hours cleaning all the tubing ends and connection areas prior to welding. Just before

welding a particular joint, I spray the tubes in the weld area with acetone and wipe off the acetone with a clean, white, lint-free cloth. I also wipe the welding rod with acetone just before I use it to weld with.

Yes, if I get in a hurry and don't take time to clean the 4130 steel tubing before welding, I can still make a strong weld, but it takes quite a bit longer to make the weld because I really have to watch out for porosity, slag and cracks in the weld. Lack of fusion also becomes a problem if I don't clean the parts prior to welding.

For parts that are not easy to sandpaper clean, try the small angle sander shown on page 20. For larger parts, such as angle steel for trailer building, use a 4" angle grinder.

Sandblasting

The dirtiest and least desirable way to clean parts before and after welding is to open-air sandblast them. But there are times when you have no other choice. For instance, you may want to modify an old boat trailer, or repair an old auto frame, or you may want to use up some steel or aluminum material that has been out in the weather for many months. The only way to clean large parts is to sandblast them.

For the times when this becomes necessary, I lay down a large sheet of plastic in an area where it will be easy to contain the sand when I am sandblasting. For instance, when I restored a 1956 Ford pickup, I sandblasted the frame in the 8-foot wide area between my garage and the 6-foot high side fence. Covering the area with a big sheet of plastic made it much easier to clean up the sand after the rust removal job was complete.

Air Compressor—You will need a minimum 1-horsepower air compressor for sandblasting. A 2- or 3-horsepower compressor will work even better. And after the sandblasting is done, you can blow off the frame/assembly with the air nozzle. The air compressor will also be necessary for operating one of the glass bead cabinets shown on the next page . If you don't own an air compressor, you can usually rent one for a day or two and save making such a large investment.

Power Sanding

It is very tempting to just sand the thin-wall tubing on a power sanding disc or on a sanding belt, but don't do it. Power sanders are designed to remove a lot of material quickly, and you will find that your .049" wall tubing is only .010" thick on the ends where you power sanded it. Unfortunately, hand sanding is the only safe way to prepare thin tubing for welding, brazing, or soldering.

Wire Brushes

I do keep an assortment of small stainless steel bristle wire brushes for cleaning the welds between passes. But I never mix the metals that I use them on. For instance, one brush is marked "Stainless Steel Only," one brush is marked "Aluminum Only," and one brush is marked "Steel Only." That is because small amounts of one metal could stick to the wire brush bristles and be transferred to a dissimilar kind of metal and cause defects in the weld. For instance, I certainly do not want aluminum, copper, brass, lead, or magnesium in my 4130 steel weld structure. The very small amounts of dissimilar metal would cause cracks in the chrome moly weld.

Don't try to remove oxidation from aluminum

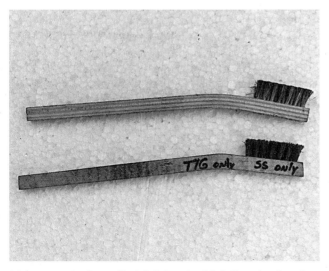

I keep a set of small stainless steel bristle wire brushes for cleaning parts before I weld them and while I am welding them. Notice that one of the brushes is marked "S.S. Only."

on a motor-driven wire wheel. The wire wheel will erode the aluminum and it will also transfer steel into the aluminum and contaminate your weld.

The same thing is true when trying to wire brush clean mill scale off 4130 steel tubing. The wire brush will erode the steel and it will imbed wire bristle material into the chrome moly and contaminate it. Again, sandpaper is the best way to clean aluminum, stainless steel and 4130 steel.

Glass Bead Cleaning

Cleaning parts for welding by glass bead cleaning can be a mixed bag. Yes, the glass beads will clean the parts with little or no erosion of the metal. However, if you are cleaning tubing, the glass gets inside the tubing and is hard to remove. On general principles, you don't want your tubular structure/assembly to be full of little, tiny beads of glass.

Several welders I have worked with would glass bead clean the weld scale off their parts between weld passes, but the glass dust must be rinsed off with solvent after the glass bead cleaning, then the solvent must be rinsed off with acetone to remove the solvent oils.

The glass bead cabinet is a very good way to clean parts after welding, in preparation for painting. But again, be sure to blow and rinse off all the glass bead residue so that it will not contaminate your paint.

Glass bead cabinets come in many sizes and price ranges. I bought a $99 special cabinet, built a wooden stand for it, and attached my shop vacuum cleaner to it to remove the dust and to evacuate the air pressure so the suction nozzle would work efficiently. It works great for occasional cleaning jobs.

If you use a glass bead cabinet on a daily basis,

Siphon-type sandblaster holds about 40 pounds of sand. It can also use other abrasives such as walnut shells. Homemade screen is for sifting sandblaster sand for reuse. After the third sandblasting, silica 30 sand becomes to fine and will no longer clean parts adequately.

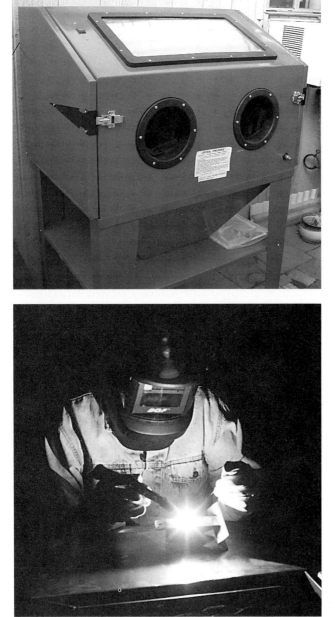

This sand blast/glass bead and walnut shell cabinet was purchased from Harbor Freight for $300 and it will do most cleaning jobs in the shop environment. It must have a shop vacuum attached to suck out the excess air and dust from the pressure blasting process.

When I was growing up on a farm, I thought all welds were eventually supposed to break because little thought was given to pre-weld cleaning and preparation. Today I know better. Good welds depend on how clean the metal and filler rod are before you weld. Don't skip this step.

you seriously need a professional-sized cabinet with quality features. These come in various sizes to suit your needs. In the airplane repair business, I have even seen walk-in glass bead cabinets where the person cleaning parts had to wear a pressurized suit to prevent glass beads and dust from getting into his breathing and vision apparatus. He looked like a deep-sea diver in a sealed suit.

Chemical Cleaning

There are several metal-cleaning chemicals on the market, but you may have to go to a paint store or to a swimming pool supply outlet to find them.

• Rust can easily be removed from steel and cast iron by applying phosphoric acid (manganese phospolene) to the metal and then rinsing with clear water. Phosphoric acid diluted comes in many forms. Auto paint stores sell it as "metal conditioner" or "metal prep." House paint stores sell it as manganese phospolene—OSHPO. And it really works. If you have never tried liquid rust removal, you will be amazed at how well it works.

•Naval Jelly stops rust, but it will not float rust away like water-thinned phosphoric acid.

•Another liquid metal cleaner is swimming pool acid, actually muriatic acid with a small amount of hydrochloric acid in it. You can buy this acid in hardware stores and, of course, in swimming pool supply stores.

•I keep a spray-bottle with a mixture of metal etch, phosphoric acid, or muratic acid to spray on parts to be welded. I also have a gallon-size glass jar with this rust remover so that I can dip small parts in it for pre-weld cleaning. Usually, 30 minutes in the acid will dissolve light rust, and a water rinse prepares the parts for welding.

Acetone, Denatured Alcohol—Although acetone is highly flammable, I keep a 1-quart can of it where I can clean the weld area and the welding rod just prior to welding. Of course, due care must be taken to prevent sparks and flame from contacting the acetone while I am welding.

The wisest thing to do is to clean the parts and the welding rod in an area outside of the welding shop and then immediately before welding, bring the parts into the weld shop.

Use Discretion

Of course, it is not necessary to clean the pieces of a utility trailer to the same level of cleanliness that you would clean your 4130 steel tubing for an airplane engine mount. But for any parts worth welding, it is wise to clean them as if you were preparing the parts for painting. Remove all rust, grease, mill scale and dirt, to the level of the intended use of the weld project.

Mill Scale

Most hot rolled, hot formed steel will have a thick layer of mill scale on it. Even the most expensive aircraft-quality 4130 chrome moly tubing and 4130 steel plate will have this thin layer of mill scale on it. Mill scale will not fuse into a weld, and it must be removed to prevent it from weakening the weld. Use sandpaper, emery cloth or abrasive screen to remove mill scale from tubing. Use sheet sandpaper or a small air angle sander and a flexible pad to remove mill scale from 4130 and mild steel sheet, plate, and angle.

Aluminum Oxide

Aluminum tubing and sheet comes in two finishes, bare and Alclad. Bare is used when anodizing is planned. Clad is more common and the cladding is actually a thin film of oxide that is allowed to form on the aluminum to prevent further corrosion.

The cladding is not weldable, and it will contaminate the weld and weaken it. Mechanical removal is recommended, including sanding and burnishing with Scotchbrite abrasive pads. Further cleaning by metal etch or phosphoric acid with a water rinse is recommended.

The best policy is: Clean the parts before you weld them as clean as you would if you were painting them.

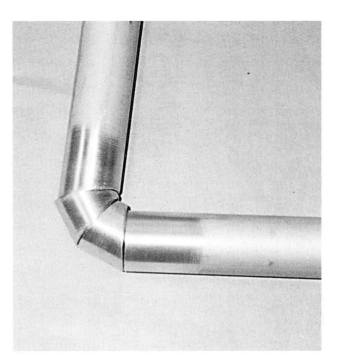

Here you see the clean, shiny areas on the tubing where the Alclad was removed to prevent it from contaminating the TIG weld.

The end result: good, high-quality TIG welds on 6061-T6 tubing that was properly cleaned.

Chapter 8
GAS WELDING AND HEAT FORMING

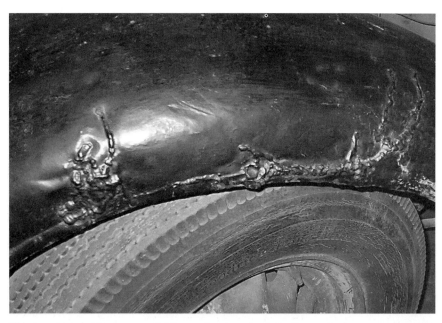

This is what welders once did to car fenders back in the early days, say in 1940. The old welders thought that the best way to weld cracks in fenders was with coat hangars and gas welding. Not pretty and not strong either. That is why the fenders continued to crack after they were welded.

This chapter tells you how to set your shop up for gas welding, and then how to gas weld chrome moly 4130 steel, mild steel, and aluminum. You will also learn how to gas-weld stainless steel. The important thing about this chapter on gas welding is that when you learn the techniques for metal fusion in gas welding, you have acquired the techniques that are basic to most other forms of welding.

Learning to gas weld or becoming better at it will provide you with welding talents that will benefit you in most of the other types of welding that you may do in the future.

When you become good at gas welding, you will be able to control the heat and control the weld metal puddle. You will be able to fuse the metal parts together so that they are actually one permanent part.

EQUIPMENT CHOICES

In Chapter 2, we lightly covered oxyacetylene welding processes, and in Chapter 3, we explained the various kinds of welding equipment you might want to use to do your welding. Again, if I could only have one welder, it would be an oxyacetylene rig with several welding tips, a cutting torch with at least two sizes of cutting tips, and at least one heating or rosebud tip. Several companies make excellent starter kits that are light and portable but also effective in accomplishing some very professional welding projects. The only drawback to the small portable units is that the gas bottles don't hold very many cubic feet of gas. You may have to refill the bottles before you finish your project.

Gas Bottle Size

However, I find that the medium-sized oxygen and acetylene tanks that hold approximately 125 cubic feet each will last through many automotive and airplane projects. You may find that a portable plastic gas welding kit and a pair of 110- or 120-cubic feet tanks will last you at least for the completion of your project, and that refilling the tanks once a year is what you will expect.

If you find later that you are using your oxyacetylene cutting torch a lot, and that you use a lot of oxygen compared to the amount of acetylene you use, you might want to use a larger oxygen tank. If you do mostly neutral flame welding, brazing and soldering with your torch, you will empty both equal-sized tanks at about the same rate.

Lease, Rent or Buy?

Probably the number one difficulty that I have experienced in all the years I have welded is in leasing, buying, or renting oxygen, acetylene and argon tanks. At the time this edition is being written, I have two sets of tanks in my welding workshop because no local welding supply dealer will refill my two 99-year-lease tanks. They won't fill my tanks because they didn't get the business of leasing the tanks to me originally. So I had to rent a pair of oxyacetylene tanks from the nearest welding supply dealer in order to continue my projects. And I have a pair of tanks sitting empty in my workshop.

After surveying several welding supply dealers from the USA West Coast to the East Coast and back again, here is my recommendation on tanks:

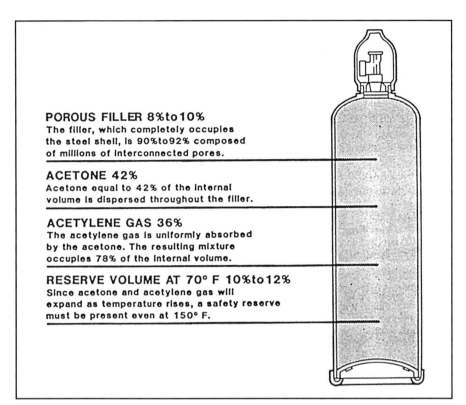

POROUS FILLER 8% to 10%
The filler, which completely occupies the steel shell, is 90% to 92% composed of millions of interconnected pores.

ACETONE 42%
Acetone equal to 42% of the internal volume is dispersed throughout the filler.

ACETYLENE GAS 36%
The acetylene gas is uniformly absorbed by the acetone. The resulting mixture occupies 78% of the internal volume.

RESERVE VOLUME AT 70° F 10% to 12%
Since acetone and acetylene gas will expand as temperature rises, a safety reserve must be present even at 150° F.

Acetylene is a compound of carbon and hydrogen (C_2H_2). It is a versatile industrial fuel gas used in cutting, heating, welding, brazing, soldering, flame hardening, metallizing, and stress-relieving applications. It is produced when calcium carbide is submerged in water or from petrochemical processes. The gas from the acetylene generator is then compressed into cylinders or fed into piping systems. Acetylene becomes unstable when compressed in its gaseous state above 15 PSIG. Therefore, it cannot be stored in a hollow cylinder under high pressure the way oxygen, for example, is stored. Acetylene cylinders are filled with a porous material creating, in effect, a "solid" as opposed to a "hollow" cylinder. The porous filling is then saturated with liquid acetone. When acetylene is pumped into the cylinder, it is absorbed by the liquid acetone throughout the porous filling. It is held in a stable condition. Filling acetylene cylinders is a delicate process requiring special equipment and training. Acetylene cylinders must, therefore, be refilled only by authorized gas distributors. Acetylene cylinders must never be transfilled. Courtesy Victor Welding Co.

CAUTION

Don't accept tanks that are out of hydro-test date or nearly due for hydro test. The hydro test fee is usually $25 or more for most high pressure tanks. Each time you accept a tank from a dealer who fills or exchanges tanks, check the due date. Most tanks are hydro tested for 10 years at a stretch.

First Choice—Buy a pair of tanks from the nearest welding dealer. Make sure that he will refill your tanks for you when they are empty. Try to get him to buy the tanks back from you if you move to another area or if you hear that he is moving.

Second Choice—Rent a pair of welding tanks from the local welding equipment dealer, but make sure he will agree to refill your tanks when they are empty.

Regulators

A single-stage regulator drops cylinder pressures from up to 2,200 psi to 2–3 psi in one stage. Most small gas-torch kits come with single-stage regulators. The biggest problem with single-stage regulators is that they allow outlet pressure to drop as inlet pressure drops. Also, regulated pressure changes with temperature. Higher temperature raises pressure and vice versa. Therefore, you must keep your eye on the regulator gauges to maintain the desired outlet pressure.

A two-stage regulator automatically reduces cylinder pressure of 2,200 psi down to about 50 psi. Pressure is adjustable down to 1–15 psi. Many large gas torches come with two-stage regulators because they use gas much faster—cylinder pressures are likely to drop rapidly due to a high gas-flow rate.

Although it would be nice if a small torch were available with two-stage regulators, many race cars or certified airplanes have been welded with small torches fed by single-stage regulators. You can

This portable gas welding torch set comes without bottles so you can rent or lease bottles from your local welding supply dealer, a less troublesome solution to the bottle problem. Photo: Harris Calorific/Lincoln Welding.

DECIPHERING THE CODE OF HYDRO-TEST DATES STAMPED ON GAS CYLINDERS.

An example of hydro-test dates is shown on the CO_2 bottle that was found in Alberta, Canada. The bottle/cylinder was first placed in service in December 1931, because the first stamp is 12-36, then 5 years later, the next date due is August 1942, then 5 years later the date is October 1947, and on and on and on. Apparently, the cylinder was in constant use during all those years since 1931 and is still in use.

Some of the newer cylinders are spun steel, and are good for 10 years in between hydro-test times. They will have a stamp that also has a star stamped in the date code, indicating 10 years rather than the older 5 years between tests. Note that the first date on the cylinder in the picture has a German Swastika stamp before the first date (arrow), that indicates this was a cylinder first used in Germany prior to WWII. Photo courtesy Roy Hamm, Edmonton, Alberta, Canada.

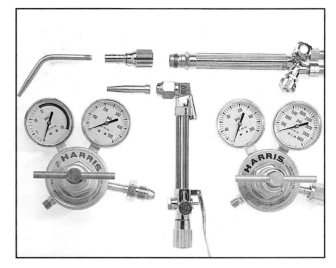

This gas welding and cutting set is a good setup for a beginner as well as a professional fabrication shop. You can buy your own gas hose in a length that suits your needs. Photo: Lincoln/Harris Calorific.

always buy a set of two-stage regulators later if you absolutely must have the best equipment.

Regulators represent about 75% of the cost of a complete gas-welding starter kit. This doesn't include the cost of the cylinders.

Accessories

Even though you may plan to buy a gas-welding kit with several welding tips, you'll still need a few extra things to make welding easier.

Torch Lighter—You must never light a torch with an open flame. Get a torch lighter or striker. Flint-type lighters make sparks similar to the way you would strike a match, and will work until the flint striker wears down. More expensive, electrical-discharge torch lighters will last many years.

Long-Handle Wire Brush—Use this brush to clean rust and welding scale from parts before welding. Weld seams must be clean and rust-free for complete fusion of base metal and filler metal.

Stainless-Steel Wire Brush—Use this small brush to clean welding scale while welding. I keep such a brush in my back pocket so it will be handy for cleaning off scale that develops during rest periods.

Welding Cylinder Wrench—This wrench is handy for removing and

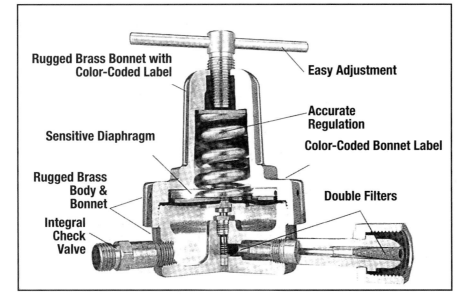

This cut-away of an oxygen regulator gives you the picture of how the adjustment knob pushes on the spring to give pressure settings from zero psi to 200 psi. Courtesy Smith's Equipment.

Rugged Brass Bonnet with Color-Coded Label

Easy Adjustment

Accurate Regulation

Color-Coded Bonnet Label

Sensitive Diaphragm

Rugged Brass Body & Bonnet

Double Filters

Integral Check Valve

Even these smallest oxygen and acetylene cylinders have hydrotest dates. The small oxygen cylinder can be filled at the welding supply store, from a larger oxygen cylinder, but remember that the acetylene cylinder must take up to 7 hours to refill for safety reasons. Expect to have to exchange the acetylene cylinder.

I keep these tools in my gas-welding toolbox. From left to right: wire brush, stainless-steel wire brush, pliers, gas cylinder wrench, machinist hammer, flint striker, lightweight leather gloves, single-lens goggles, soapstone, temperature-indicating crayon, tip cleaners, rattail file, half-round file and flat file.

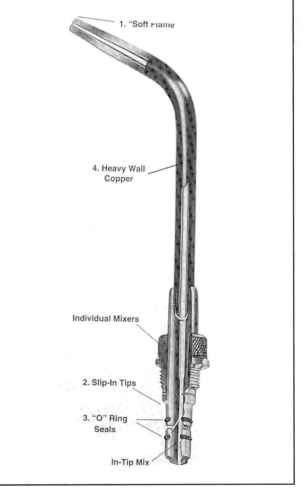

1. "Soft Flame

4. Heavy Wall Copper

Individual Mixers

2. Slip-In Tips

3. "O" Ring Seals

In-Tip Mix

Cutaway of a Smith's torch tip, showing the intricate parts necessary to mix a good welding flame. Photo: Smith's Equipment.

replacing the pressure regulators while changing cylinders. Also, this wrench can be used for opening and closing the acetylene- or hydrogen-cylinder valve if it doesn't have its own knob

Pliers—Use pliers to pick up hot pieces of metal just welded. You can also use them to hold pieces in place while you tack-weld them.

Small Machinist Hammer—Often, you'll need a small hammer to bend hot metal into place, or tap a part into place before continuing the weld.

Safety Glasses—These clear-lens glasses are an absolute must for a welder's toolbox. Don't confuse them with welding goggles. Wear safety glasses to protect your eyes whenever chipping, filing, grinding or sawing metal. Welding goggles reduce available light too much.

Leather Gloves—Wear leather gloves to shield your hands from welding heat. They allow you to weld for longer periods of time without the need to stop to cool your hands. Because they'll burn, never pick up hot metal with leather gloves. Use pliers instead.

Soapstone Marker—This chalk-like marker doesn't burn off until the metal melts. Use it for making reference marks on metal or to mark lines for a cutting torch.

Temperature Indicators—Temperature-indicating crayons and paint are convenient for determining the temperature of metal for heating or forming. See page 6.

Weld-Tip Cleaner—As with paint brushes, gas-welding and gas-cutting tips must be cleaned. Cleaning the outside of a tip is easy, but cleaning the inside requires a spiral tip-cleaning rod. The cleaner comes with a variety of rod sizes, each matched to a specific tip size. Each rod has a

This cutaway of a gas welding/cutting torch body gives you an idea of why we say that your gas welding rig is considered delicate equipment. Handle it carefully and it will give you long, dependable service. Photo: Smith's Equipment.

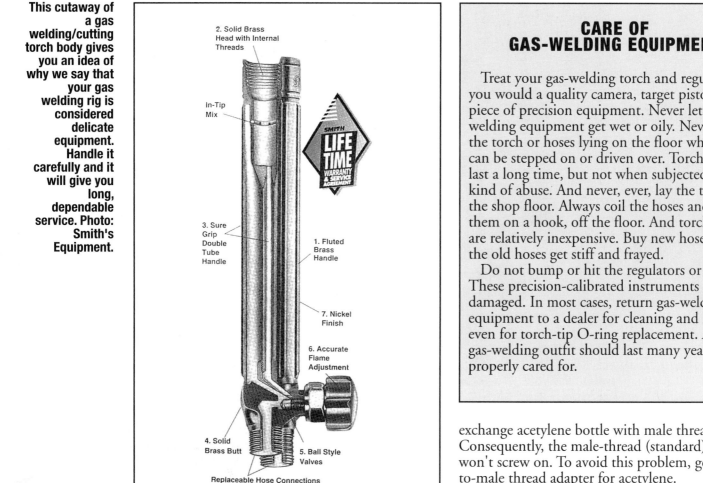

2. Solid Brass Head with Internal Threads

In-Tip Mix

3. Sure Grip Double Tube Handle

1. Fluted Brass Handle

7. Nickel Finish

6. Accurate Flame Adjustment

4. Solid Brass Butt

5. Ball Style Valves

Replaceable Hose Connections

If your torch doesn't have safety check valves, or flashback arrestors, install one between torch and ease hose (arrows).

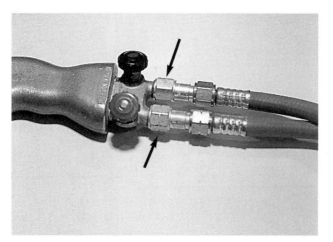

CARE OF GAS-WELDING EQUIPMENT

Treat your gas-welding torch and regulators as you would a quality camera, target pistol or any piece of precision equipment. Never let gas-welding equipment get wet or oily. Never leave the torch or hoses lying on the floor where they can be stepped on or driven over. Torch hoses last a long time, but not when subjected to that kind of abuse. And never, ever, lay the torch on the shop floor. Always coil the hoses and hang them on a hook, off the floor. And torch hoses are relatively inexpensive. Buy new hoses when the old hoses get stiff and frayed.

Do not bump or hit the regulators or gauges. These precision-calibrated instruments could be damaged. In most cases, return gas-welding equipment to a dealer for cleaning and repairs—even for torch-tip O-ring replacement. A quality gas-welding outfit should last many years when properly cared for.

precision fit in its tip hole. Use the tip cleaner as you would a rifle-bore rod.

Metal Files—I use three different files in my welding toolbox: a coarse round file, a coarse half-round file, and a flat mill file. Files are used to fit parts before welding.

Acetylene Regulator Adapter—You might get an exchange acetylene bottle with male threads. Consequently, the male-thread (standard) regulator won't screw on. To avoid this problem, get a male-to-male thread adapter for acetylene.

Welding Cart—Instead of buying a welding cart, build your own. It's relatively easy and you'll gain valuable experience doing it. I built mine many years ago, and it's still in use. You need training and practice before you start a complicated welding project. So, what better way to gain experience and proficiency than building something simple, but useful? Welding cart plans are provided on page 139.

To make the cart, you'll need to use many of the procedures described in this book: fitting, cutting, butt welding, corner welding, T-welding and brazing. Read the sections that apply to each.

BASIC WELDING PROCEDURES

It is a very good idea to gather up a lot of scrap pieces of the kind of metal you plan to use in your projects. If you plan to build an experimental airplane, buy several scrap pieces of 4130 steel tubing from one of the aircraft supply houses. All of the tubing supply companies sell welders practice kits for just a few dollars.

If you plan to build race cars or even trailers, go to your local steel and aluminum supply dealer and

GAS WELDING TIPS, SIZES AND GAS FLOW DATA CHART

Decimal Metal Thickness	Fractions Metal Thickness	Victor Tip Size	Smith's Tip Size	Henrob Dillon Tip Size	Drill Size	Oxygen Pressure (psig)	Acetylene Pressure (psig)
.015"	1/64" to 1/32"	000	AW200	No Band .017	.020	3	3
.030"	1/32" to 3/64"	00	AW201	—	.025	3	3
.070"	1/32" to 5/64"	0	AW203	1 Band	.035	3	3
.090"	3/64" to 3/32"	1	AW204	—	.040	3	3
.125"	1/16" to 1/8"	2	AW205	2 Bands	.046	4	4
.190"	1/8" to 3/16"	3	AW207	—	.060	4	4
.250"	3/16" to 1/4"	4	AW209	—	.073	4	4
.375"	1/4" to 1/2"	5	AW210	—	.090*	5	5

*For metal thickness over .375", 3/8", 9.5mm, use arc welding.

CYLINDER SIZES

Size	Cu. Ft.	Oxygen Height	Wt. Full	Size	Acetylene Cu. Ft.	Height	Wt. Full
R	20	14"	14 lbs.	MC	8	14"	8 lbs.
AXL	58	41"	54 lbs.	B	33	23"	26 lbs.
Q	92	35"	70 lbs.	2AWQ	55	31"	61 lbs.
D	125	48"	124 lbs.	#4	90-150	36"	113 lbs.
S	155	51"	92 lbs.	#4	151–230	37-1/2"	150 lbs.
K	251	56"	153 lbs.	#5 WK	250–380	43-1/2"	20 lbs.
H	281	56"	162 lbs.				
T	337	60"	172 lbs.				

buy several pounds of cut-offs. You can also go to local welding shops and ask to go through their scrap bins for metal to do your practice welding on. Then, of course, the welding projects in Chapter 16 of this book will provide you with some very good practice welding on things that do not have to be perfect at first.

My first serious practical gas-welding project was welding a go-kart. After several dozen go-karts, I progressed to welding bird-cage frames for sports racing cars. Long after that, I decided to attend welding classes at the local community college.

Prepare Your "Coupons"

The welding trade calls sample welds and practice welds "coupons." This chapter shows you drawings and photos of several kinds of welding coupons for gas welding. Do a good job of fitting and preparation (cleaning) on your coupons. You might as well have good metal to practice on: dirty, poorly fitted metal will not teach you much.

Get Organized First—Make sure your weld area is well-organized and that there are no flammables in the shop. This includes paint cans, oil cans, rags, wood, and for sure, no gasoline cans. Make sure that your clothes are not easily flammable. Wear a heavy denim shirt and pants and cotton socks, and good, tight-fitting shoes. Do not wear synthetic fabrics such as polyester and nylon blends, as these will melt to your skin if burned. Wearing a welder's cap is advised also. Soft leather gloves that allow good hand movement should be worn. But don't

SAFETY TIPS FOR GAS WELDING

• Never tilt acetylene cylinders on the side when in use. The acetone stabilizer will flow into the regulator and damage it.
• Mark full cylinders FULL with a marking pen. Mark empty cylinders EMPTY with a marking pen.
• Store cylinders at less than 125°F (52°C). Make sure valves are closed and caps are on stored cylinders.
• Chain or otherwise restrain all cylinders in an upright position.
• Always crack—slightly open—the valve to blow out dust before attaching a regulator. This helps prevent contamination of the regulator, which may cause erroneous gauge readings.
• Never haul cylinders in the closed trunk of a car because of the explosion hazard from escaping gases.
• Wear goggles with the correct filter lens.
• Don't wear oily or greasy clothes when welding.
• Wear leather or denim clothes when welding; leather is best. Blue jeans are great.
• Don't cut or weld material coated with zinc, lead, cadmium or galvanized coating. Poisonous fumes are generated as the coating burns off. Coated sheet steel is used in an increasing percentage of late-model cars and trucks, especially in rust-prone areas such as rocker panels, fenders, rear quarters, door outers, cowl plenums and trunk floors.
• Never use oil or grease on gas-welding equipment. It's extremely flammable at high temperatures, particularly in the presence of oxygen.
• Leak-test hoses, regulator and torch before lighting. Use soap solution or non-oily leak-detector solution on the connectors and look for bubbles. Never use an open flame to check for leaks.
• Oxygen fittings have right-hand threads. Hose is green or black.
• Acetylene fittings have left-hand threads. Grooves around flats on nuts identify left-hand threads. Hose is red.
• Never adjust an acetylene regulator to more than 14 psi. More pressure and flow will draw acetone from the cylinder. Without acetone, acetylene is very unstable and could explode.
• Keep pliers handy to pick up parts you've just welded or cut so you don't burn your gloves or fingers.
• To minimize soot when lighting an acetylene flame, add more acetylene. Don't use equal amounts of acetylene and oxygen. The resulting loud pop may cause an accident. Be prepared to add oxygen when the flame ignites. Never try to light the torch with a small amount of acetylene. This causes tremendous amounts of soot.

Adjust oxygen regulator to about 3 psi on low-pressure gauge, left. Low-pressure acetylene regulator should also be adjusted to 3 psi.

wear heavy leather gloves, because you will not be able to handle the torch and welding rod with heavy gloves.

Welding Table

If you have not yet built your own portable welding table, you can simply place the 3/8"-thick steel plate that the plans on page 137 call for on top of a portable stool or other suitable platform, and put at least two firebricks on the steel plate to insulate your coupons or the torch flame from the steel plate.

Open the Cylinder Valve

There really is a correct procedure for opening the oxygen and the acetylene cylinder valves. Each valve is to be opened slowly and carefully to avoid damaging the diaphragms in the regulator. Open the acetylene tank valve about 3/4 to one turn and no more. This is to prevent too much accidental flow of acetylene as might happen if the valve were fully open and if you cranked in the adjustment past 15 psi. If the acetylene tank flow exceeds 1/7 of the tank volume per hour, the acetylene could become unstable and cause the tank to explode.

Also, if you have an emergency and need to shut off the (fuel) acetylene quickly, 3/4 to one turn open makes it easier to shut off the gas.

Next, slowly open the oxygen valve, but open it all the way, several turns to make the valve seal set properly. The oxygen tank valve could leak if the valve is in the mid-open position. Again, if you open the high-pressure oxygen tank valve too quickly, the pressure surge could damage the oxygen regulator diaphragm. Now you are ready to

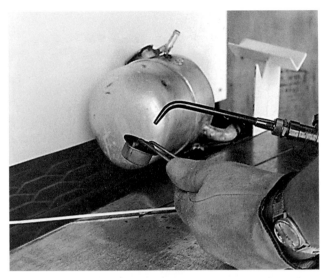

Open torch acetylene valve about 1/2 to 3/4 turn. Hold striker under torch tip and light torch. Remember, NEVER LIGHT AN ACETYLENE TORCH WITH AN OPEN FLAME!

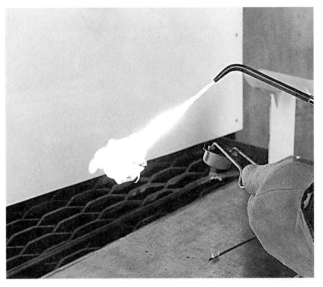

Acetylene-only flame should look like this. Quickly add oxygen to eliminate soot.

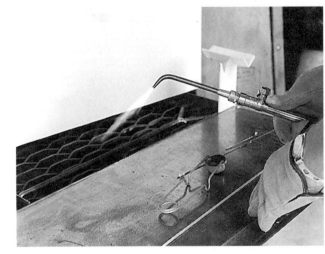

As you add oxygen, three distinct flame cones will appear. Continue adding oxygen until you have a neutral flame. It looks like this with #3 Victor tip, 3 psi oxygen pressure and 3 psi acetylene pressure. One distinct cone is at center with light blue outer flame.

light the torch, to adjust the flame, and to practice gas welding.

SAFE PROCEDURES FOR LIGHTING & SHUTTING OFF OXYACETYLENE

For many years, welders were taught to light the torch with both oxygen and acetylene flowing, and to shut off the acetylene fuel first. But this procedure often caused popping and even flashback accidents, so a new procedure has been initiated to prevent gas welding explosions.

Lighting the Torch

1. Adjust the acetylene regulator to hold a predetermined pressure with the torch acetylene valve opened 1/2 turn. Then shut off the valve.
2. Adjust the oxygen regulator to hold a predetermined pressure with the torch oxygen valve opened 1/2 turn. Then shut off the valve.
3. Open the acetylene valve 1/8 turn and light the torch with the striker. Some black soot will come out, so increase the acetylene flow by opening the torch valve until the flame gets bigger and the smoking stops.
4. Next, open the torch oxygen valve about 1/8 to 1/4 turn and adjust for a neutral flame.

Shutting Off the Torch

1. Close the oxygen valve first. Oxygen supports combustion and aids explosions, so get rid of the oxygen first.
2. Next, shut off the acetylene valve.
3. If you intend to light the torch again within a few minutes, you need not bleed the lines and

regulators.
4. When you are through welding or heating or cutting for even a few hours, close the oxygen and acetylene tank valves and bleed the pressure from the regulators, hoses and torch by opening each valve separately.
5. When the pressure is bled off, close the torch valves and back out the regulator adjustment nut until no drag is felt on the screw. Backing out the adjustment screws relieves pressure on the adjustment springs and reduces strain on the regulator.

Every time your oxygen-acetylene torch pops, carbon and sometimes flame will flow back into the torch body and will eventually cause torch burn-out and possibly a torch explosion. However, I have never personally experienced a torch explosion in all the years I've been welding.

WELDING TECHNIQUES

Now it is time to light up the torch and to begin

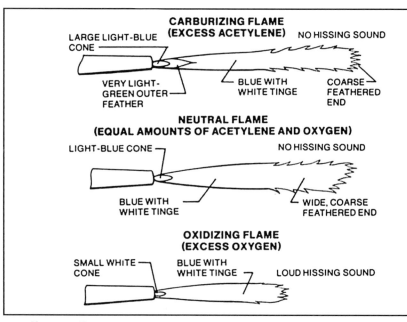

CARBURIZING FLAME (EXCESS ACETYLENE)

LARGE LIGHT-BLUE CONE — NO HISSING SOUND — VERY LIGHT-GREEN OUTER FEATHER — BLUE WITH WHITE TINGE — COARSE FEATHERED END

NEUTRAL FLAME (EQUAL AMOUNTS OF ACETYLENE AND OXYGEN)

LIGHT-BLUE CONE — NO HISSING SOUND — BLUE WITH WHITE TINGE — WIDE, COARSE FEATHERED END

OXIDIZING FLAME (EXCESS OXYGEN)

SMALL WHITE CONE — BLUE WITH WHITE TINGE — LOUD HISSING SOUND

Here are the basic gas-welding flames. Each has a distinctive shape, color and sound. Sounds vary from soft (low gas flow) to medium (medium gas flow) to loud (full gas flow) at specific pressures. Neutral flame is used most.

practice welding projects. Adjust the oxygen and acetylene regulators for 4 psi each and put a .040" (#1) tip on the torch handle. You are ready to light up and to adjust for a neutral flame. Next you can start practicing on a piece of steel.

Making a Puddle

The first thing to do is make a molten puddle on the steel plate. With your welding goggles on and over your eyes, direct the neutral flame at the steel. Oscillate the torch tip in a half-moon, zig-zag or circular pattern as shown. The exact pattern is not important. The idea is to keep moving the torch in a rhythmic, repeatable pattern as you move the weld puddle along, but without overheating it. Use the pattern that works best for you.

Working Distance

Hold the torch about 1" from the work. The steel should start to turn red within 5–10 seconds. If it doesn't you're not holding the torch close enough, or the tip you selected isn't large enough for the metal thickness.

I usually tell first-time welding students to overheat the metal first and melt holes in it. This is because the normal tendency is to weld too cold, resulting in poor penetration. So go ahead and burn a few holes until you can control the heat and maintain a puddle.

Learn to manipulate the torch by moving it closer to the metal, then backing up quickly if a hole starts to form. After experimenting with the puddle for about 10 minutes, you're ready to advance to the next step.

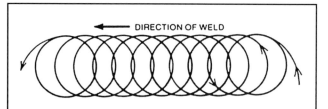

DIRECTION OF WELD

Use circular motion to preheat base metal before forming puddle. Circular motion spreads heat evenly. When steel turns dull red, you are ready to make molten puddle.

CAUTION: NEVER FLOW MORE THAN 1/7 BY VOLUME OF ACETYLENE CYLINDER PER HOUR

Because of the unstable nature of acetylene, the bottle could overheat and explode.

If more heat is required, for instance when preheating or maintaining heat for brazing a large cast iron structure, it would be advisable to use 2 or more acetylene tanks to provide the 1/7 volume per hour of heat flow. Also, consider 2 or 3 extra helpers, each operating a separate, independent oxyacetylene heating torch to provide the required heat. Think of the heating requirements in units, BTUs. If one torch only provides 1/2 the necessary heat in BTUs, then use 1 or 2 extra torches to bring up the BTUs.

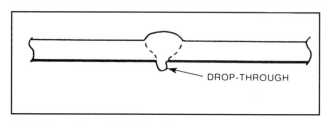

DROP-THROUGH

Slight drop-through is OK as it ensures 100% penetration. Reduce heat to reduce or eliminate drop-through.

Running a Weld Bead

The next step is to practice running a bead on your piece of scrap steel. Start by making another molten puddle. With your left hand—if right-handed—momentarily dip the welding rod into the puddle, then withdraw it. If you're left-handed the torch goes in the left hand and the rod in the right. Always dip the rod into the molten puddle. Never try to heat the rod and puddle together. If you do, the flame will melt or even vaporize the small-diameter rod before the base metal gets hot enough to puddle. Remember, form a puddle, then intermittently dip the rod into the puddle to add

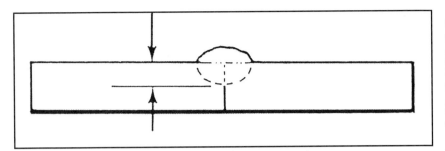

Good weld must have sufficient penetration, or depth of fusion. Weld has 50% penetration. Increase heat or slow down to increase penetration.

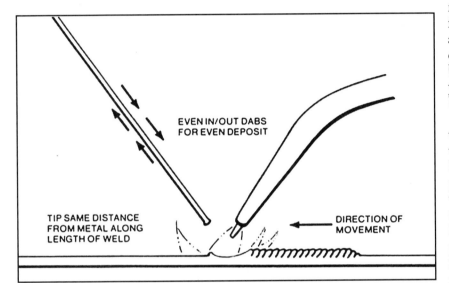

EVEN IN/OUT DABS FOR EVEN DEPOSIT

TIP SAME DISTANCE FROM METAL ALONG LENGTH OF WELD

DIRECTION OF MOVEMENT

Forehand welding: It takes coordination to use filler rod. Deposit an even amount of filler rod with each dab as you move molten puddle along weld seam. Practice will develop rhythm. Drawing by Ron Fournier.

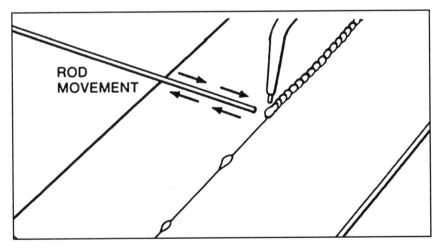

ROD MOVEMENT

Sheet metal needs plenty of tacks—about every 1"—to reduce warpage. Tack welds are melted into weld bead as final bead is made. Drawing by Ron Fournier.

filler material as you run the bead.

You've seen those beautiful welds that look like a row of fish scales? Well, they resulted from a welder doing the dip, dip, dip thing. If the welding rod sticks in the puddle, point the flame at it, melt it off and try again. Sticking is caused by not dipping fast enough or not keeping the puddle molten. Just keep practicing.

Forehand Welding—Forehand or backhand welding refers to the direction you point the torch tip in relation to the direction you're running the weld bead. If you're forehand welding, the torch is angled so it points in the direction of the weld. This is to preheat the base metal so it puddles easily as you move along with the weld bead.

Backhand Welding—Like walking backward, backhand welding is similar to welding backward. The technique is to point the torch at the already-welded seam, away from the unwelded seam. This prevents the base metal from being preheated—usually an undesirable feature. Backhand welding is rarely used, except to avoid burning through very thin metal. The added mass of weld bead may help absorb the extra heat. However, so does pulling the torch away from the work.

Tack Weld—As previously discussed, a tack weld is nothing more than a very short weld that's used for holding two pieces in place prior to final welding. You'll make a lot of tack welds.

Stitch Weld—A stitch weld is used where a continuous weld bead would be too costly and time-consuming, and where maximum strength is not required. Although it can vary with the application, a stitch weld typically is made up of short weld beads about 3/4" long, spaced by equal gaps.

Butt Weld—Once you've mastered the art of running a bead, you're ready to try welding two pieces of metal together. Let's start with a basic joint. A butt weld is a weld made between two pieces laying

Typical stitch welds: Technique is used where continuous weld is not required for strength, and it is also used to avoid warpage on long sections. The stitch weld procedure can also be used for Arc, TIG and MIG welds.

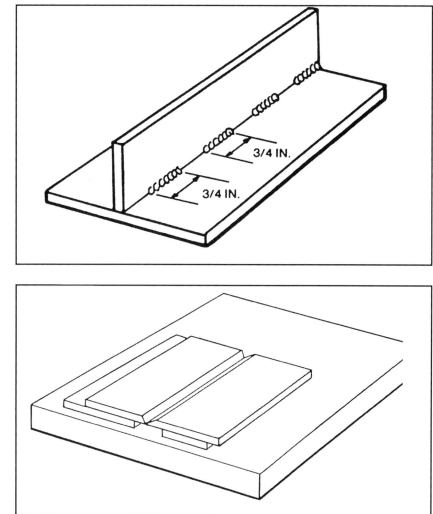

3/4 IN.

3/4 IN.

Space pieces to be butt welded off table on scrap pieces or firebrick so you won't contaminate weld, waste heat or damage tabletop.

alongside and butted against one another, edge to edge or end to end.

Place two pieces of metal side-by-side and butt them together. There should be no gap. The seam will be welded into one solid bead.

PRACTICE, PRACTICE

By now, you should have collected some scraps of steel to practice welding on. Ideal for welding practice would be several 2 x 5-in. pieces of 0.032–0.060-in. thick mild steel. They don't have to be exactly this size, but your practice work will look better if the pieces are uniform.

To avoid contaminating the weld bead with firebrick or the welding table, raise the metal pieces off the table or brick by inserting extra pieces of scrap underneath each workpiece, but not under the weld seam.

Next, tack-weld the two pieces together, first at each end, then about 1 in. apart, along the length of the seam. This keeps the metal aligned during the welding. The trick here is to keep both edges at

the same temperature by manipulating the torch. Add a little heat until the puddle forms, then dip the rod in two or three times until you have a good tack weld.

After tack-welding, use your pliers to hold the work so you can check for warpage at the weld seam. Straighten the pieces by tapping on one piece with a hammer when you hold the other with the pliers. You're now ready to run a solid bead. If you're right-handed, do a forehand weld by starting at the right end of the seam, make a puddle—torch in right hand—dip the rod and keep going until you get to the left end. If you're left-handed, reverse hands and start welding at the left end of the seam.

As you come to each tack weld, remelt it into the puddle. When complete, you won't be able to see the tack welds; they'll be part of the weld bead.

Test Weld

Because the appearance of a weld can fool the beginner, test each weld for soundness. "Pretty" welds can literally break in two if there is insufficient penetration—not enough filler fused with the base metal. The weld bead may only be "laying" on the base metal. Ideal penetration can be from as low as 15%—weld bead is fused into the base metal by 15% of overall thickness—to over 100%—it's fused the full thickness of the base metal and sagging through to the back side. Clamp the piece to be tested in a vise, just below the weld seam. With a big hammer, bend the top piece toward the top of the weld bead. Chances are that the weld will break through the back side, perhaps completely if penetration is poor. A common cause of broken welds made by beginning welders is crystallization. Crystallization is caused by excessive gas pressure—usually, too much oxygen pressure.

It is much better to weld with 1–2 psi gas pressure for any size tip and avoid overheating the weld. If your weld breaks, don't give up. Look for lack of penetration. With your next test weld, try to get a good puddle going before you dip in the filler rod. Practice makes better welds.

LEAK TESTING

It's always a good idea to leak-test gas-welding equipment from time to time. Do this both after you've set up brand-new equipment or are using equipment with which you're unfamiliar. Why? Aside from the obvious safety considerations of flammable gas leaks, leakage can cause fluctuating pressure, upsetting torch mixture settings.

Use a soap solution, such as liquid dish soap. Never use a bare flame or oil. Oil is a combustible hydrocarbon, so even a little on the oxygen-hose connection could lead to disaster.

To do a leak test, turn one regulator screw fully counterclockwise. Open the cylinder valve. Make sure the torch valves are closed. Build up pressure in the regulator and hose by slowly turning in the regulator screw—turn it clockwise. This should raise line pressure from about 5 to 15 psi. Now, apply the soap solution to each connection and look for bubbles.

Leak-test both cylinders and their regulators, hoses and the torch.

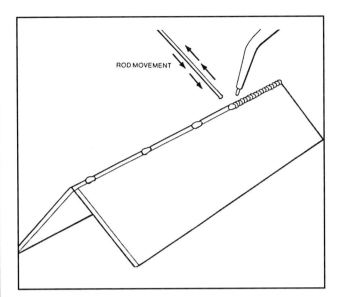

ROD MOVEMENT

Outside corner weld requires less heat than inside-corner weld. For maximum penetration, metal edges should not overlap, but form a V. Drawing by Ron Fournier.

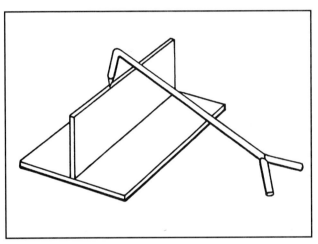

When making T-weld, support vertical piece with mechanical finger. Tack-weld each end of joint. Point flame at horizontal piece 70% of welding time, but manipulate torch to put equal heat on both pieces as you run bead.

Fillet Weld

A *fillet*, or *outside corner*, weld is a weld performed on two pieces of metal joined in a V-type configuration. The weld bead is run on the outside rather than in the Vs inside, or "crotch." An outside-corner weld is easier to make than a butt weld because it takes less heat to maintain a puddle and run the weld bead. It's easier to get the edges hot because they're up in the air and only the edges are being heated. Less heat is lost to supporting members, such as the steel worktable.

Block the two pieces of metal up like the peaked roof of a house. You can use a heavy piece of metal at the sides to hold the two pieces in place for tack welding. When you do this, you have just created your first welding jig!

Tack-weld the two ends, then make some tack welds in between. Now, run a continuous weld bead. A near-perfect corner weld is one with slight penetration through to the underside—100% penetration of the base metal.

T-Weld

T-welds are the most difficult of all the practice welds. That's why we waited until last to try it.

When making a T-weld, you're making another type of fillet weld—welding in a corner where two pieces join at 90° or so.

Block up two pieces of scrap metal with a metal finger as shown in the accompanying illustration. Viewed from the end, the two pieces form an upside-down T. Block up the flat piece in the area of the weld so the welding table won't absorb the heat of the weld. Firebricks or short sections of angle stock are great for this.

Next, tack-weld the two ends as before. Then make more tack welds about 1 in. apart along the weld seam. Again, if you are right-handed, start welding from the right end—vice versa for you southpaws. Direct most of the heat to the horizontal piece and less to the vertical—at about a 2-to-1 ratio.

The reason for directing more heat at the horizontal piece is that the weld is being made in its center, so there's more volume of metal to absorb heat. There's only half as much volume in the vertical piece because you're welding its edge.

Plug weld is made to help join tubes, one slipped inside another. Note angle-cut–scarfed–end of larger tube to increase weld-bead length. This weld is also called a rosette weld.

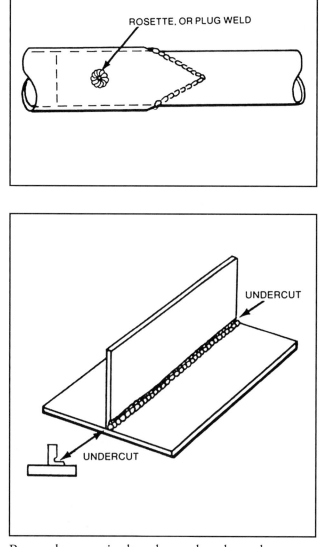

ROSETTE, OR PLUG WELD

Problem with making T-weld is undercutting vertical workpiece and insufficient penetration on horizontal workpiece. Undercutting can be eliminated by correct manipulation of torch and rod. If your weld looks like this, point flame at undercut and add filler rod as puddle forms there.

UNDERCUT

UNDERCUT

Remember, manipulate the torch to keep the puddle going on both pieces, and feed the rod into the puddle by intermittent dipping.

A common problem with T-welds is ending up with an otherwise decent weld that has an undercut or gouge in the vertical piece. This is caused by failing to adequately manipulate the torch to keep equal puddles on both pieces and overheating the vertical piece. The molten base metal from the vertical piece runs down and solidifies into the weld bead.

The solution is to tilt the torch away from the vertical piece when you see the undercut start, and at the same time, dip the rod into the undercut part of the puddle. You'll have to do a little torch twisting, but that's what it takes—that and a lot of practice.

Remember, control the temperature and the puddle, and the weld will take care of itself.

GAS WELDING TORCH SOUNDS

Different volumes of oxygen and acetylene can be fed through a specific-sized welding tip to produce varying amounts of heat. For instance, a Victor or Smiths welding tip size no. 1, or AW204, can be used to solder (250°F), braze (800°F) or fusion-weld steel (3,000°F). I have even used a heavy-duty cutting torch to solder 1/2" copper tubing to a cold drink aluminum can.

The amount of oxygen and acetylene flowing through the welding tip will provide varying BTUs of heat to the weld. Here is how I adjust the gas flows for 3 different heat requirements with a #1 welding tip:

1. Soft flame, for soldering, 4 psi oxygen and 4 psi acetylene pressures: Adjust for a very quiet flame sound, just barely a whisper of flame voice from the flame. Solder should flow in 2 to 3 seconds after applying the flame.

2. Medium flame, for brazing, 4 psi oxygen, 4 psi acetylene pressures: Adjust for a stronger flame sound. The sound will be like the sound of water running from your aerated bath lavatory faucet while you are washing your hands.

3. Maximum flame, for fusion welding, should have a very strong sound, obviously putting out a good amount of heat. It could also be described as a hissing sound. Use this flame for welding thicker metals, over .080" thick, and for heating parts for bending.

COMMON PROBLEMS
Torch Pop

This will scare the dickens out of you and blow sparks all over the place! Torch pop is ignition of the gases inside the torch. It is more common when welding in corners such as when making a T-weld. Torch pop is caused by an overheated tip too close to the metal. This causes a small explosion inside the tip. After the torch pops several times, the tip gets dirty from the metal splattered on it, causing it to pop even when you aren't too close.

The solution for preventing torch pop caused by a dirty tip is to shut off the torch and clean the tip with your tip cleaner. Light the torch and try welding again. If cleaning the tip didn't cure the problem, increase oxygen and acetylene pressures 1 psi from the previous setting and adjust for a hotter

Using the welder's mechanical finger to hold two parts in place for welding, use the fire brick to insulate the parts from the cold welding table.

CORRECT GAS PRESSURES

As an EAA Technical Counselor, I am often asked to observe gas-welding and cutting practices by experimental aircraft builders. The most common mistake I see is too much gas pressure, especially too much oxygen pressure. About 3 to 4 pounds oxygen and acetylene pressure is about right for most fusion welding of thin-wall tubing and other steel parts that are less than .080" thick.

It seems that most self-taught welders found that 10 pounds acetylene and 20 pounds or more oxygen pressure worked fairly well for cutting torch work, so they just use the same pressures for gas welding. (The proper pressures for cutting torch are 3 to 4 psi acetylene and 20 to 25 psi oxygen to cut 1/4" to 3/8" steel.)

Too much gas pressure, especially too much oxygen pressure, will burn up and crystallize your welds. After all, in cutting torch work, you use 20 to 25 pounds oxygen pressure to OXIDIZE and VAPORIZE the steel you are cutting.

Check the chart on page 59 for correct gas pressures to use in gas welding.

flame. If this doesn't cure the popping problem, try the next larger tip.

Flaky Welds/Poor Penetration

Such welds break apart when you bend them. They're caused by not making the puddle hot enough before dipping the rod. You just can't melt rod and drop it onto the base metal, hoping it will stick. Instead, it must become a homogeneous part of the base metal by mixing, or fusing, while in the molten stage.

The solution is to get the weld puddle hotter. Do this by using a larger tip or by holding the torch closer to the work, but not so close that torch pop results. Remember, if you can't make a molten puddle in 5–10 seconds, use a larger tip. This is why I tell beginner welders to melt holes in the metal, if necessary, but get it hot! Usually, it takes no more than 15 minutes of extra practice and you can be making good welds with this technique.

Rod Sticks to Base Metal

Every beginner experiences rod-sticking problems because the weld puddle isn't hot enough. The solution is simply to heat and maintain a molten puddle. The puddle melts the rod, not the torch. If you keep a good puddle going, then merely dip the rod where you want filler material. You'll get a good weld bead. Concentrate on the puddle, and the weld will take care of itself.

Flashback

This is a potentially dangerous condition where the gas burns back through the torch and hose to the regulator and cylinders, damaging the torch, hose and regulator. The cylinders are next, and an explosion is possible!

Flashback is usually accompanied by a loud hiss or squeal. If it occurs, flashback must not be allowed to continue! Immediately shut off oxygen at the tank if flashback occurs, then shut off acetylene. The oxygen is first because it supports combustion. Flashback is usually caused by a clogged torch barrel or mixture passage. Don't relight the torch until you cure the problem.

For these reasons, every oxyacetylene torch should be equipped with flashback safety arrestors. Basically, these are one-way valves that install in the torch gas lines.

Solutions

Don't be afraid to move the torch as necessary. You have 6,300° F (3,482°C) available at the tip of a gas torch. Position the tip close enough to a piece of steel that melts at 2,750° F (1,510°C), and the metal will melt!

If the puddle gets too big, pull the torch away for a second or two to give the metal a chance to cool and solidify. If you burn holes on one side of the weld seam and are not getting the other side red, direct the torch away from the hot side and toward the cold side to get more even heat.

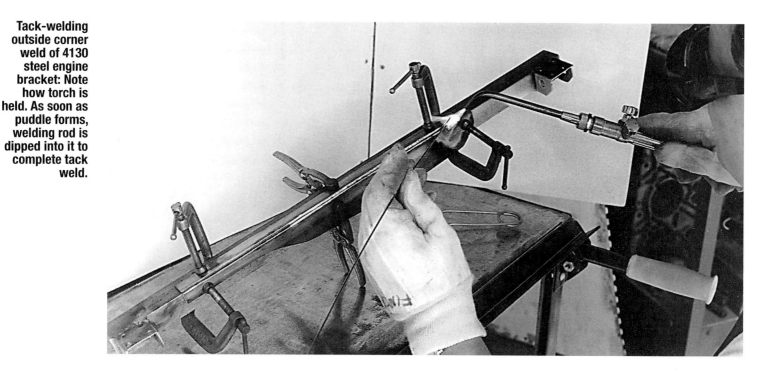

Tack-welding outside corner weld of 4130 steel engine bracket: Note how torch is held. As soon as puddle forms, welding rod is dipped into it to complete tack weld.

You are probably getting tired of hearing this, but temperature control is the key. Control the temperature and you control the weld puddle. After selecting the correct tip size, temperature is controlled by the direction of the torch, the distance the tip is from the work and by gas-pressure settings. Follow these simple rules:

• Point the torch where you want the heat.
• Aim the torch away from where you don't want the heat.
• Back away if the puddle is too hot.
• Move closer if you are not getting the puddle hot enough.
• Increase heat by opening the torch valves if you can't get enough heat.
• Decrease heat by closing the torch valves if the puddle is too hot.
• Move the torch in an oscillating pattern.

GAS-WELDING SAE 4130 STEEL

After you've built your gas-welding table, and maybe another useful project or two, you should be ready to try welding chrome moly, or SAE 4130, steel. SAE 4130 welds about the same as mild steel, but it's more likely to become air-hardened and brittle from improper welding. Nevertheless, don't be afraid of 4130. If you can weld a nice bead with mild steel, you can learn to do the same with 4130.

Don't use copper-coated rod for welding 4130 steel. It may cause cracks and bubbles in the weld. Use only bare mild-steel or bare 4130 rod. For most jobs, 1/16-in. diameter rod is the best size to use. It comes in 36-in. lengths. I cut them in half for better control—did you ever try writing with a yard-long pencil?

Never braze 4130 steel. Its woodlike grain will open up and let brass flow into it. When the brass solidifies, the steel will then have thousands of little wedges that cause cracks between the grains. Sometimes the cracks will propagate as you watch!

Cleaning—Keep 4130 tubing or sheet clean of all oil, rust and dust. Clean it before you weld it. Don't even touch the weld area with your fingers after cleaning. Use methylethyl ketone (MEK), acetone or alcohol to clean both the base metal and welding rod. You can't get it too clean!

File, sand or sandblast all scale from previous welds before welding over them. The scale could contaminate your weld if not removed.

Shop Area

Even though it is generally good practice to keep the welding area clean, well-lit and draft-free, it is especially important for welding 4130. A bright, clean shop area helps you make clean welds. A dark, dirty welding shop will contribute to cracks, pinholes and generally poor welds. Before you actually start building a long-term project, go out to the welding shop and write someone a letter in the position you'll be welding in! If you're not comfortable writing the letter, you certainly won't be comfortable welding. And never allow any drafts of air, cold or hot. One welding instructor once

Two seams are shown here in aluminum that I am gas welding. The seam on the left is 100% penetration, but I have overheated and melted a big hole in the seam on the right. Too much heat is bad, too little is bad, too. Learn to CONTROL the heat.

Typical aircraft fuselage gas welding. Next step is to wipe the tubes down with metal prep liquid to remove the rust film, and then prime with rust preventative primer.

advised me to not even let my dog wag his tail in the welding shop!

Weld Technique

Evenly preheat the weld area to about 375°F (190°C). Although preheating to the precise temperature is not critical, a temperature-indicating crayon or paint can be used for getting the feel for how hot this is. Play the flame—move it back and forth—over the entire weld seam, holding the torch tip about 4 in. from the metal.

Start welding where a minimum of preheating is required to form a puddle—such as on the edges. After running a bead for a fraction of an inch or so, the metal is automatically preheated, particularly if you're using the forehand method. This saves preheating time and reduces the chance of overheating the weld.

If you tack-welded the seam prior to running the final weld, be sure to remelt the tack welds along with the base metal and include them in the weld as you come to each.

Never jerk the torch away as you complete a weld. Hydrogen and oxygen in the air will contaminate the weld and it will cool too rapidly, possibly cracking it. After finishing a weld, pull the torch back slowly. Let the weld cool to a dull red before removing the torch completely. Not only does this reduce the chance of cracking the weld, pulling the torch back slowly also allows the molecules to relax gradually and stress-relieve (page 73) somewhat. Even when stopping for more welding rod, hold the torch 4 in. from the work so the flame "bathes" the weld in heat.

Never weld the back side of a 4130 weld unless the designer specified it. If welded properly, the joint will be strong enough without doing so. Besides, the back side of a weld probably has scale that should be sandblasted prior to welding.

When redoing cracked welds on 4130 steel, file or saw out any bad welds and start over. You might even have to put a patch plate over the joint if excess metal was removed. A patch plate is usually made of the same material and thickness as that being patched. Extend the patch plate 200% past the damaged area and weld the plate all the way around.

Drill a relief hole in tubing that's being welded closed. If you don't, air pressure building up from the heat inside the tube will blow out the last of your weld as you finish sealing the tube. Therefore, drill a #40 or 3/32" hole in a non-stressed area about 1" from the end of every tube to be welded shut. If you want, squirt spray preservative such as LPS-or WD-40 into this hole. Or you can leave the tube dry as I recommend, and either weld the hole shut or seal it with a Pop rivet. If you rivet the hole shut, coat the rivet with sealer to keep out moisture.

Rust Prevention

Rarely is it necessary to add oil to preserve the inside of a 4130-steel tubular structure. If moisture can't get inside the tubes, they won't rust. Most rust occurs from outside. Paint will protect it there. Oil is heavy, messy, and may contain chemicals harmful to 4130. I've repaired rusty fuselages from airplanes built in the '30s, and the rust was on the outside, not inside. There was no oil preservative inside the tubing.

GAS WELDING ALUMINUM

Most people equate oxyacetylene welding with gas welding. That's because acetylene is by far the

ALUMINUM WELDING Q&A

A handy question-and-answer pamphlet for welding aluminum was prepared for the annual Experimental Aircraft Association (EAA) Fly-in at Oshkosh, Wisconsin. This pamphlet was based on the most commonly asked questions concerning aluminum welding by the 200,000 EAA members. With their permission, portions of the pamphlet follow.

What aluminum alloys are weldable?
Answer: 1100, 3003, 3004, 5050, 5052, 6061 and 6063 are weldable. Specifically, 1100 is dead soft and not good for structures; 5052 is medium hard and good for fuel tanks; and 6061-0 is soft, but can be heat-treated after welding to make it very hard and strong. 3003, 3004, 5050 and 6063 are weldable, but seldom used. You can weld 3003, 3004, 5050 and 6063, but don't use these for your project. Stick with 5052 and 6061 for better results.

What kind of rod should be used?
Answer: 1100 rod for 1100 material and 4043 rod for all other alloys. Flux-cored rods work well also.

What flux should be used?
Answer: Antiborax #5 for cast aluminum and #8 for sheet aluminum.
Note: There are many other fine aluminum-welding fluxes.

At what temperature does aluminum melt?
Answer: Pure aluminum melts at 1,217°F (658°C)—less than half that of steel—but alloys melt at lower temperatures. Aluminum oxide, a corrosive film that forms on aluminum immediately after cleaning, melts at a much higher temperature than aluminum. This oxide must be removed before welding and inhibited during welding. Remove oxide with a stainless-steel wire brush, Scotchbrite abrasive pad or acid. Use flux before and during welding to prevent oxide formation.

What equipment is needed to weld aluminum with oxyhydrogen?
Answer: You need a standard gas-welding torch, one oxygen regulator and cylinder, another oxygen regulator converted for use on a hydrogen cylinder and a cylinder of hydrogen. You should also use cobalt-blue lenses in your welding goggles. With the conventional green lenses, all you would see of the flame and weld puddle would be a large yellow spot. The blue lens filters out the yellow light blocking your view.

What size welding tip should be used for welding aluminum with oxyhydrogen?
Answer: Use a tip three times larger than the one used for welding 4130 steel of the same thickness. For example, if you would use a #1 tip for welding 4130 steel, use a #4 tip for welding aluminum of the same thickness.

most popular fuel used for gas welding. But when it comes to gas-welding aluminum, hydrogen is often recommended instead of acetylene. If you do use hydrogen in place of acetylene for welding aluminum, be sure to switch back to acetylene to weld steel. Welding steel with oxyhydrogen will cause hydrogen embrittlement—hydrogen contamination of the weld joint—causing it to be brittle.

Oxyhydrogen

For many years, aluminum welding was a mystery to me. I thought TIG was the only way to weld aluminum. Then I bought some aluminum welding flux, bare aluminum rod and tried gas welding aluminum. It surely is different from gas-welding steel, but it works!

Oxyhydrogen is the preferred method of welding aluminum. This is because its 4,000°F (2,204°C) neutral-flame temperature is closer to the 1,271°F (658°C) melting temperature of aluminum than is the 6,300°F (3,482°C) flame temperature of oxyacetylene.

Aluminum vs. Steel

When using oxyhydrogen to weld aluminum, remember that an oxyhydrogen flame has little color. It doesn't look like an oxyacetylene flame. Therefore, don't try to adjust it to the same color. It will have two distinct cones—inner and outer—as described for oxyacetylene welding. When properly adjusted, oxyhydrogen flame has an almost clear outer cone and pale-blue inner cone. However, the adjusting procedure is the same.

Use hydrogen just as you would acetylene. Almost all the procedures for lighting the torch, choosing tip size, tip-to-work distance and other techniques for welding aluminum are essentially

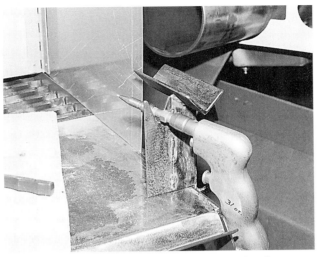

Make a weld bench hangar for your gas torch so you won't have to shut it off and relight each time you stop to adjust or setup your weld. A Dillon torch is shown here.

The VHT. ceramic paint-coated exhaust headers on my aircraft conversion Buick V6 engine were gas welded. The engine mount was TIG welded.

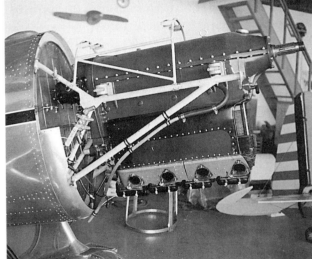

Here is a 1930s aircraft engine mount that was gas welded. Nearly 80 years later it is still good welding, and it is working.

the same as for welding steel. Temperature control through practice is still the secret, regardless of the material or method.

One major difference in welding aluminum compared to steel is that the puddle is much cooler than steel and will not melt the welding rod as easily. So you must preheat the welding rod slightly by holding it near the puddle and partly in the flame as you move the puddle along. The dip, dip, dip process is the same.

Another difference with welding aluminum vs. steel is that there's little change in base-metal color as it melts to form a puddle. Aluminum doesn't change color and glow red like steel when it's heated. Instead, it stays the same color until a few degrees before reaching its molten state. At this instant, it becomes shiny where the puddle forms. If you continue heating the aluminum puddle past this state, looking for a color change, the puddle will drop out. You'll have a hole rather than a puddle.

Always use a welding rod with a diameter closest to base-metal thickness. If you have a metal shear and you're welding sheet stock, filler rod can be made by shearing square strips from the sheet. This will automatically give you the precise filler-rod size and alloy.

For best results, the aluminum pieces must be absolutely clean. You should even wipe the welding rod with a clean, white cloth before welding.

Aluminum welding rod must be used with flux. Otherwise, you won't be able to weld aluminum successfully. Flux must be used to remove and inhibit the harmful oxides that form on aluminum. Mix aluminum welding flux powder with water or

alcohol. While welding, frequently dip the welding rod into the mixture to keep the rod coated with fresh flux. Brush on the flux every two or three minutes. The base metal must have fresh flux on it while welding.

Keep a large bucket of fresh, clean water nearby. After cooling, dip the parts in the water to rinse off the flux. Flux left on aluminum causes corrosion.

HEATING & FORMING TIPS

When you heat steel to dark blood red (1,050°F or 566°C), it bends easier than when cold. Heat steel to bright cherry red (1,735°F or 746°C) and you won't believe how easily it bends! Steel molecules become more "plastic," or pliable, the hotter they get.

You can use heat to assist in bending large, thick pieces of steel with a simple bench vise and a small

ROSEBUD TIPS—HEATING WITH ACETYLENE

Victor Torch No.	Smith's Torch No.	Acetylene Pressure	Oxygen Pressure
4	AT603	6–10	8–12
6	AT605	8–12	10–15
8	MT603	10–14	20–30
10	MT605	12–14	30–40
12	MT610	12–14	50–60
15	—	12–14	50–60

Other fuel gases that can be used for heating include propane, natural gas, propylene and MAPP gas.

pry bar. I modify new trailer hitches to fit old cars simply by heating and bending them. I learned that trick when I was about seven years old, helping my uncle make new plow tips for his farm tractor. Red-hot steel bends like warm taffy candy.

Practice bending hot steel and pretty soon you'll be doing things you never thought possible. Just use the appropriate-sized welding tip, cutting tip or rosebud tip to apply the amount of heat needed. I try to use a tip that heats the metal to cherry red in less than one minute. If it takes longer, the tip is too small. This wastes both time and gas.

Rosebud Tip

A rosebud tip is so called because its flame configuration looks like a rosebud. Strictly meant for heating, this tip is actually useful for both heating and forming metal. A rosebud tip is no good for welding, but it does generate gobs of heat. It also uses a lot of oxygen and acetylene, so make sure your tanks aren't low before starting a project.

The first time I lit a rosebud tip, it produced a pop that sounded like a shotgun! The next six times I did it, I got the same loud noise! Then it occurred to me that 4 psi oxygen and acetylene pressure were insufficient to operate that big tip! I then increased oxygen pressure to 25 psi and acetylene pressure to

10 psi; the torch lit with only a soft pop. It takes a lot of pressure to operate a rosebud tip!

Rosebud tips are good for freeing stuck parts. However, because of their high-heat output, use a temperature-indicating crayon or paint to check the temperature so you don't overheat the part.

Cutting-Torch Tip

If necessary, you can use a cutting-torch tip for heating. It puts out more heat than a welding tip, but less than a rosebud tip. Adjust the torch for the maximum neutral flame without the cutting lever while you're heating a part, or you'll oxidize or cut it!

AUTOBODY GAS WELDING

From the first steel-bodied cars until about 1980, gas welding was the only way to repair severely damaged car bodies. After 1980, an increasing number of car bodies use high-strength steel (HSS) panels. Depending on the alloy, many high-strength steels have a crystalline grain structure that can be destroyed if heated above a certain temperature. The limit for martensitic steel, for example, is 700°F (371°C). This severely weakens the metal and can cause cracking.

For this reason, auto manufacturers recommend that HSS panels be welded with other techniques—MIG welding is one of them. But for all those older cars, you can make good use of your 6,300°F (3,482°C) gas welder for body-and-fender repair.

To determine whether the body panel on a late-model automobile or light-duty truck is HSS, consult the shop manual. Because each manufacturer uses different types of HSS and may have different repair procedures for even the same alloy, follow its procedures to the letter. Many HSS panels are critical structural members and the repair procedures are difficult, so don't be surprised if the manufacturer recommends replacing a damaged HSS panel rather than repairing it.

STRESS-RELIEVING MYTHS

For at least 70 years, old-time welders have advised new welders to reheat tubular structures after the structure has been completely welded. They tell us that the purpose of reheating the welded cluster to dull red is to "stress- relieve" the weld area.

The truth is that open-air, manual application of oxyacetylene flame to a weld cluster is likely to do more damage to the structure than to simply leave it alone after welding. I will describe the correct procedure for stress-relieving a 4130 steel tube aircraft engine mount, and you be the judge of whether it is possible to accomplish the safe, correct procedure in your own workshop situation:

1. To prevent the stresses from distorting the mount, it should be firmly bolted to a very heavy fixture, and the fixture should be a heat-resistant stainless alloy. To prevent stresses in the fixture, it should be bolted and not welded together. The fixture to hold a 15-pound engine mount would weigh more than 100 pounds.

2. The engine mount and fixture would be placed in an air-tight oven and brought up to 1,150°F to 1,175°F temperature and held at that temperature for 2 hours. The color of the ENTIRE engine mount will be DARK CHERRY RED, and it will hold that color for 2 hours. Two hours are required to allow the entire mount to undergo a complete residual stress relief process.

3. After holding the mount and stress-relief fixture in the oven for 2 hours, the mount is removed from the oven and allowed to cool in still air (not wind, and positively not in water or oil). The complete cooling process generally takes 4 hours.

4. After the mount and fixture cools to 100°F or less, the mount is removed from the jig. Comments: It should be obvious that this correct stress-relieving process could not be accomplished in a welding shop.

• Heating the mount to 1,575°F, bright red and scaling, will anneal the mount, soften it, and cause it to lose strength, possibly back to less tensile strength than mild steel.
• Expert heat treaters, who work to mil-specs, agree that 4130 tubular steel weldments, as found in engine mounts, when properly fitted and cleaned before welding, and properly welded, will never stress crack in normal use. Furthermore, if no cracks exist immediately after welding, the likelihood of cracking long after welding is extremely remote.
• Magnaflux or dye penetrant checks for cracks and porosity after welding is really sufficient. You probably should check the engine mount for residual magnetism that might exist after MIG or TIG welding. Demagnetize it to prevent compass distractions.

Chapter 9
TORCH CUTTING

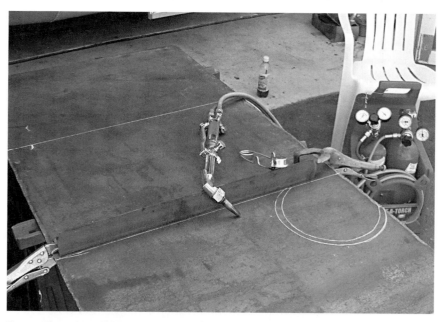

For cutting a straight line on this sheet of 3/16" steel plate, a piece of angle steel was clamped to the plate with vise grip pliers and the cutting torch was lit and the straight cut was made. Note the small portable oxyacetylene unit that works great for small jobs.

O ld-timers in the welding trade like to call torch cutting "burning it off," and they are partly correct in using this term because oxygen-fuel cutting actually works by oxidizing the metal in the cut.

GAS CUTTING & OXIDIZING

Now that you've mastered the art of gas welding, the next item on the list is learning to flame-cut with acetylene—sometimes referred to as a blue wrench or gas wrench. These terms came from mechanics who used cutting torches for loosening or removing stubborn bolts, nuts or other seized parts, usually due to rust. Our main purpose is to use the acetylene cutting torch for fabricating steel parts.

The primary chemical reaction in flame-cutting steel is oxidation. Because cutting is really oxidizing, you cannot cut metals that do not oxidize (rust) easily, such as aluminum and stainless steel.

So primarily, you can flame-cut mild steel and cold-rolled steel. These steels make up a large part of things we use, such as automobiles, trailers, farm equipment and more. Flame-cutting is useful for cutting elaborate shapes not suitable for cutting with a bandsaw. Remember that flame-cutting leaves rough edges with slag that must be final-trimmed later. This is useful if you're doing arts and crafts, but a problem when doing precision fitting.

USING A CUTTING TORCH

After learning to gas weld, you will discover that a cutting torch is relatively easy to operate. Simply light the torch, adjust the flame to neutral, make a little puddle, push the

oxygen lever and you're cutting steel! Here are the steps for cutting 1/4" thick steel plate:

• Select a piece of 1/4" scrap steel plate. Also find a piece of angle steel similar to the one shown in the previous picture. Use it to guide your torch hand and help you cut a straight or smooth curved line. Secure it with C-clamps or locking pliers, if necessary. Mark the cut-line with soapstone.

• Position the plate so the cut-line hangs over the edge of the welding table. Or, lay two short sections of angle iron facedown on the bench and lay the plate on top. This will prevent a nice cut being made across your workbench top.

• Before you light the torch, check your clothing. You should be wearing cuffless pants, high-top shoes, a long-sleeve shirt, gloves and welding goggles. If all is OK, proceed to the next step, lighting the torch.

Light Cutting Torch

Adjust the regulators for about 4 psi acetylene and 10–15 psi oxygen. Oxygen pressure is much higher than acetylene pressure because the oxygen does most of the work.

Next, preadjust the cutting tip. Shut off the oxygen valve on the cutting tip and fully open the oxygen valve on the torch handle. This is important. Open the acetylene valve about one turn and light the torch using a flint or electric striker. Add oxygen by opening the oxygen at the cutting tip until you have a neutral flame.

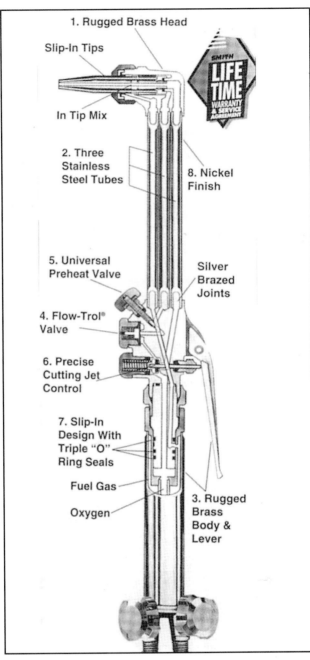

The cut-away schematic of a typical Smiths cutting torch shows its complex design. Study the cut-away to see which valves adjust the acetylene, the preheat oxygen and the cutting oxygen flows. Courtesy Smith's Equipment.

1. Rugged Brass Head
Slip-In Tips
In Tip Mix
2. Three Stainless Steel Tubes
8. Nickel Finish
5. Universal Preheat Valve
Silver Brazed Joints
4. Flow-Trol® Valve
6. Precise Cutting Jet Control
7. Slip-In Design With Triple "O" Ring Seals
Fuel Gas
Oxygen
3. Rugged Brass Body & Lever

Start Cutting

If you're right-handed, support the torch with your left hand while resting it on the plate. This will allow you to guide the torch along the cut-line for a more even cut. Grip the torch with your right hand and have your thumb ready to press the oxygen-cutting lever. Start your cut by heating the edge of the steel plate at the right end of the cut-line. Reverse everything if you're left-handed.

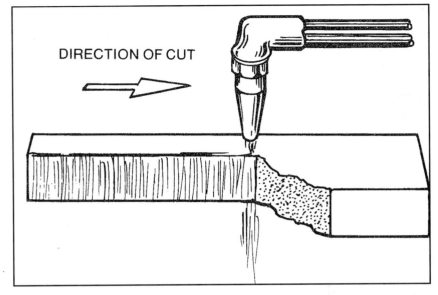

DIRECTION OF CUT

Drop cut occurs when cutting speed and oxygen flow are correctly matched so there is no drag. Flame exits kerf immediately below where it enters.

Note that when cutting with acetylene, material is removed. The resulting void is the *kerf*. Consequently, you must cut on the outside of the cut-line—scrap side. Otherwise, the part you're making will be short by the kerf width.

When a puddle develops, press on the oxygen lever. The flame should immediately begin to blow away—oxidize—the metal. If it doesn't, you didn't have a good puddle. Continue heating the metal and try again. When the sparks fly as the cut begins, carefully guide the torch along the cut-line. With a 1/4" steel plate, you move along at about 1" every three seconds, following the cut-line. Move too fast and you'll get a shower of sparks back in your face because the metal wasn't heated enough. Move too slow and you'll overheat the metal, resulting in excessive slag.

Be careful when you reach the end of your cut. A chunk of hot metal may fall to the floor if you've made a clean cut. But, chances are the slag will hold the pieces together. A light tap with a hammer should break them apart.

Tips for Better Flame Cutting

• Excess slag at bottom of cut indicates the preheat flame is too hot. Correct by reducing acetylene pressure or using a smaller tip.
• Metal doesn't have to be super clean for cutting.
• Most beginners force the cut by moving too fast. Slow down. Refer to the chart on page 79 for ideal cutting speeds for different metal thicknesses.
• Clean cutting tip periodically. The cutting process tends to splatter molten metal back on the cutting tip, reducing cutting efficiency.

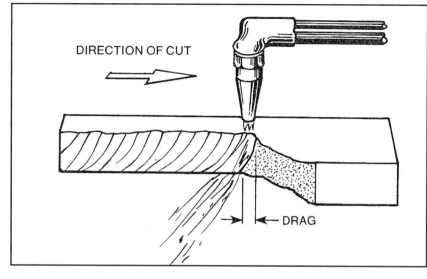

If speed is too high or oxygen flow not enough, drag occurs when cutting thick material. Reduce or eliminate drag by reducing speed or increasing oxygen flow–pressure.

The cut is started by heating a corner of the steel to red hot BEFORE squeezing the cutting trigger. If you squeeze the cutting trigger before the steel gets red, the cut will not start.

Flame-Cutting Aids

There are a few items you can use that'll make a cutting project easier. They range from a piece of angle iron used as a guide to cut a straight line to a sophisticated flame-cutting machine. Chances are you won't need the cutting machine unless you'll be duplicating several pieces from heavy steel plate.

Marking Tools—I already discussed soapstone, center punches and scribes for laying out and marking cut-lines on page 39. To assist with marking, you should make patterns. A pattern duplicates the shape you want to cut out. It can also be used for checking the fit of the final part without actually having the part.

Once you've developed the pattern, lay it on the material to be used for the final part. Then trace around the pattern with soapstone or whatever marker you desire.

Pattern material is inexpensive. All you need is plenty of thin, flexible sheets of cardboard. Such material is available at office supply stores. Corrugated cardboard from boxes is cumbersome but OK to use in a pinch. To mark patterns, you need soft-lead pencils and a pencil compass. Additional drawing equipment such as straightedges, 30°/60° and 45° triangles and felt-tip markers should also be on your list of pattern-making items. Finally, you'll need a pair of heavy-duty scissors for cutting out patterns.

Cutting Table—If you'll be doing a lot of flame cutting, it would be helpful to have a cutting table (page 137). A cutting torch can't distinguish a metal-top welding table from a workpiece. Consequently,

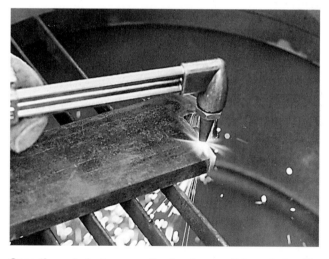

Once the cut starts, move the torch smoothly and steadily in the direction of the desired cut, holding the trigger down. To make neat, clean cuts, practice, practice, etc.

you shouldn't flame-cut anything that's lying directly on top of the table. Otherwise, you'll end up making a cut through the tabletop as well.

To avoid this, either raise the workpiece off the table by supporting it with scrap angle iron, hang it over the edge of the table or set it somewhere else. Whatever you do, don't support it with anything you don't want to be cut or damaged. Ideally, it is best to support the work with a cutting table.

A flame-cutting surface doesn't have to be exotic. It can be as simple as several sections of angle iron, positioned corners up, bridging two metal saw-type horses. Or, it can be an honest-to-goodness table.

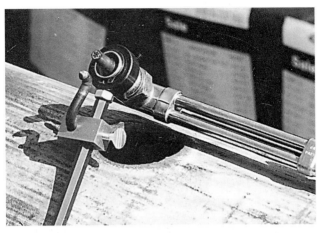

A circle cutting guide attachment can be adapted to most cutting torches for cutting circles and accurate radiuses up to 24".

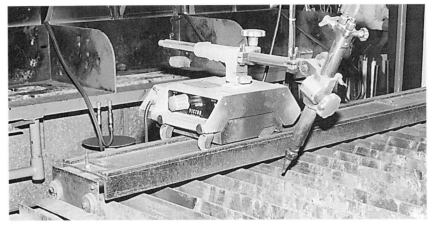

Typical welding school motor drive straight-cut oxy-fuel cutting torch. Obviously, steadiness and consistent movement of the torch makes for better cuts.

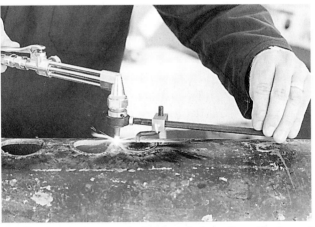

This is the proper way to hold a circle cutter attachment and cutting torch.

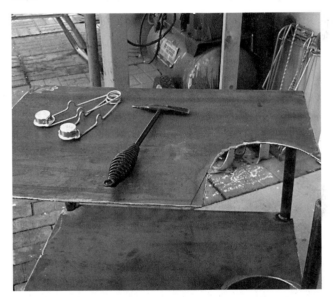

Here we have made a circular cut to allow an argon bottle / cylinder to sit on the cart for a more convinent way to store and use a new TIG welder. The cut was made with a torch and then cleaned up with a 4" angle grinder.

The table would consist of an angle-iron frame with four legs. Instead of using a solid steel top, string several sections of angle iron loosely between the frame. This will allow you to adjust the angle iron to support the work as desired. You can also replace them easily when they are cut.

Cutting Machines—Cutting machines used in the welding industry can flame-cut parts to exacting dimensions, duplicate parts or cut out many parts at one time. Some of this equipment is very expensive, such as an electric-powered machine that runs on its own track. More expensive machines have multiple cutting heads. While a tracer follows a single pattern, the multiple cutting torches cut out several duplicate parts simultaneously.

Chances are you won't need such equipment. For the hobbyist or small welding shop, inexpensive flame-cutting machines for cutting out small parts are available, such as the one from Williams' Low-Buck Tools. Manually operated with a tracer, this machine is designed to install on a 55-gallon drum. The drum contains all the sparks and scrap that would otherwise fall to the floor.

PISTOL-GRIP CUTTING TORCH

This torch was originally called the Dillon MK III after its inventor, but recently is marketed as the Henrob torch. This torch does a very good job at cutting steel plate and thin sheet steel. It has a pistol-like grip.

Cutting with a Henrob Torch

Traditional gas torches use an oxyacetylene cutting tip with six small flames in a small circle to preheat the steel, and a large orifice in the middle of the six smaller holes. The larger orifice is where the stream of oxygen comes out when you depress the lever on the torch body for cutting. This means that no matter which direction you travel with the cutting torch, you are following the nice, clean-cut

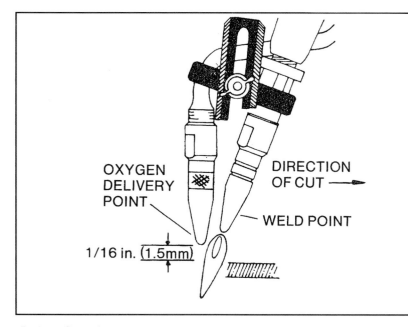

Henrob cutting torch has guide wheel assembly that clamps to torch to assist in making precise cuts. Torch must be turned to follow cut-line. Photo courtesy of K. Woods, Inc.

Preheat flame is separate from oxygen-delivery tip on Henrob cutting torch. Consequently, cut must be made with oxygen tip trailing preheat tip as shown. Oxygen tip is adjusted to other side of preheat tip for cutting sheet metal. Drawing courtesy Shannon Marketing, Inc.

kerf, with at least one or two more oxyacetylene flames that tend to melt the metal back together.

The Henrob cutting torch attachment is different. It uses a single oxyacetylene flame to preheat the steel and a single oxygen-only stream of higher-pressure gas to make the cut. No acetylene flame follows, so the cut stays clean. But you don't get something for nothing. The Henrob cutting attachment works only in one direction. If you want to cut a circle, you have to move the cutting attachment in a circle to keep the preheated oxyacetylene flame and oxygen in a leading-trailing relationship.

Cutting thin sheet metal, such as car fenders, is really where this torch excels. For sheet-metal cutting, use a different attachment that places the oxygen-only tip at the rear of the flame so you make the cut going away from you. This allows very thin sheet metal to be cut quickly and with almost no heat-affected zone adjacent to the kerf. This feature is ideal for the new high-strength steels that tend to crack in the heat-affected zone.

CUTTING STEEL WITH OXYGEN ONLY—NO FUEL!

Here's a trick I show new welding students to illustrate that cutting with oxyacetylene is primarily an oxidizing process. Try it after you learn to use the cutting torch.

Start your cut in a piece of steel plate about 1/4" thick. After establishing the cutting speed and travel, shut off the acetylene torch valve with your left hand—right hand, if you're left-handed—while you continue to cut. You'll still be able to cut the steel with oxygen only! Of course, if you don't move steadily, you'll lose the molten puddle and the oxidizing or cutting process will stop. You'll then have to relight the torch to resume the cut.

SETUPS FOR CUTTING STEEL

Material Thickness (in.)	1/8	1/4	1/2	3/4	1	1-1/2	2	4	5	6
Suggested Tip Number	00	0	1	1	2	2	2	3	3	4
Oxygen Pressure (psi)	5 to 10	10 to 15	10 to 20	15 to 25	20 to 30	25 to 35	30 to 40	35 to 45	40 to 45	45 to 50
Acetylene Pressure (psi)	1 to 2	1 to 2	2 to 3	2 to 3	2 to 4	2 to 4	2 to 4	3 to 5	3 to 5	4 to 6
Cutting Speed Per Minute (in.)	20 to 23	18 to 20	14 to 16	12 to 16	10 to 14	8 to 12	6 to 10	5 to 8	4 to 6	3 to 5
Oxygen Used Per Hour (cu. ft.)	50	80	120	140	150	170	220	350	420	500
Acetylene Used Per Hour (cu. ft.)	8	10	12	15	16	17	19	25	29	30

OXYACETYLENE CUTTING TIP SIZES, GAS FLOW DATA

Decimal Metal	Fraction Metal	Victor Tip Size No.	Smiths Size No.	Henrob Tip Size	Thousandths of inch`	Oxygen psi	Acetylene psi
.125"	1/8"	000	—	Copper 8	.020"	20	3
.250"	1/4"	00	MC-12-00	—	.025"	20	4
.375"	3/8"	0	MC-12-0	—	.035"	25	4
.500"	1/2"	1	MC-12-1	—	.040"	30	4
1.000"	1"	2	MC-12-2	—	.046"	35	4
2.000"	2"	3	—	—	—	40	5

Victor Rosebud #6 = 6 holes x .043"
Henrob Cutting Copper 1 hole x .050" 8 bands

Chapter 10
GAS BRAZING AND SOLDERING

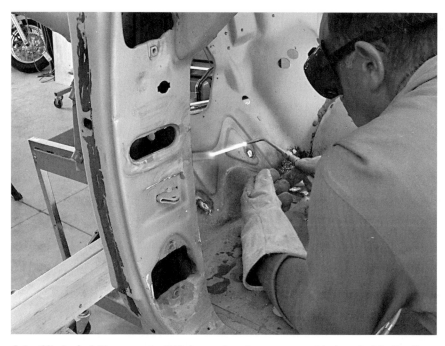

John Gilsdorf of Alamogordo, NM, is gas brazing some rust holes shut in the floor of his 1956 Ford Parklane 2 door station wagon restoration job. Brazing is faster and works just as well as cutting the floor out and splicing a new piece in.

Brazing and soldering are metal-joining methods that do not rely on melting the base metal to join two or more pieces. Instead of fusing the filler and base metals, they depend on surface adhesion of the solder or braze filler. This is made possible by capillary action—surface of the molten filler is attracted to fixed molecules nearby. When the filler metal cools, it bonds to the base-metal surface.

Brazing and soldering are more akin to adhesive bonding than welding. A major advantage of these processes over welding is that brazing and soldering are done at lower temperatures. Soldering is done below 800°F (427°C); brazing is done below the melting point of the base metal, usually below 1,500°F (816°C). Therefore, warpage and temperature-induced stress in the base metal are lower. For example, to fusion-weld a bicycle frame, you must heat the metal to the melting point of steel—more than 2,700°F (1,482°C). To braze that same frame, 1,000°F (538°C) is all that's necessary.

Another big advantage of brazing and soldering is that field repairs are simple. All you need is a torch. To arc- or TIG-weld in the field, such as repairing farm equipment, you'd have to perform the operation near an electrical outlet or a portable welder. That's not always easy to do.

In case you were wondering, brazing does not mean brass any more than soldering means lead or silver. For instance, there is brass, aluminum and silver brazing. Brazing refers to the temperature, not the metal.

BRAZING

Most people think that brazing isn't as strong as arc welding because they've seen or heard of brazed joints breaking. To disprove this, compare the tensile strengths of welding and brazing rods. E-7018 is considered to be the best arc-welding rod and it has a 70,000 psi tensile strength. Compare this to some brazing rods on page 26, which have a tensile strength of 80,000 psi or more!

To demonstrate the tensile strength of a brazed joint, I often ask my welding students to guess how much their cars weigh. The answers usually range from 2,500 to 5,000 lbs. You probably know what's coming next. I then take two 1" x 5" long, 0.060" thick mild-steel strips, overlap the ends 1" and braze them together.

Next, I put the brazed strips in the welding shop pull-test machine and let the class watch as the machine pulls over 3,000 lbs. before the base metal stretches and breaks! The brazed joint never breaks. Instead, the metal at one end of the joint fails. By this time, the whole class is saying, "That little 1 sq. in. of brazing could lift that entire 3,000-lb. car!" More importantly, the brazed joint is stronger than the metal itself.

Seam or Joint Design

The design of a seam or joint is very important in deciding whether to use solder or braze. For instance, avoid brazing or soldering butt joints. The main reason for this is the lack of sufficient wetted base-metal surface for filler to adhere to. Consequently, the base metal pulls away from the braze or solder under tension—pulling—or bending loads. Lap joints are another matter. The wetted surface area can be increased by simply increasing the overlap of the two pieces. And the joint is in shear—the best type of joint for brazing or soldering. See the illustrations of types of joints and loads, below.

Low Heat Required

Because brazing requires less heat, and therefore results in less warpage than fusion-welding, brazing is used extensively in autobody repair. Less heat also allows you to braze near rubber parts because of the reduced chance of burning the rubber.

The lower heat and induced stresses also mean that you don't have to be as careful about avoiding air currents when brazing as you do when fusion-welding sensitive metals such as 4130 steel tubing. And materials of different thicknesses can be joined easily. You don't burn up the thin part trying to heat the thick part.

Brazed joints look good and usually require only flux removal to maximize their final appearance. Brazing, by its nature, will flow into a smooth fillet, giving the joint a finished look without filing or machining.

Remember: Never fusion-weld a joint that was previously brazed. If you have a project that requires both fusion welding and brazing, fusion-weld first, then do the braze joint. Never braze first. Otherwise, you'll boil away the braze filler with the higher temperature of the fusion weld.

You can maximize the strength of a brazed joint by giving the joint more surface area. Depending on base-metal thickness, roughen it by grinding, sanding, sandblasting or coarse filing. All those surface scratches increase surface area and provide a tooth the filler can cling to.

Identify the base metal and use the correct filler metal. For instance, if you tried to braze shiny stainless steel with shiny aluminum filler rod, the metals just would not mix. You'd end up

John is brass brazing a rust hole shut while restoring his '56 Ford station wagon. He is using a bare brass rod and dipping it in the can of brazing flux near his right elbow. He heats the rod with the torch, dips the rod into the flux and then brazes some more on the floor. This method works OK for areas that only have a few rust pinholes.

with a hot mess that will fall apart.

Always heat the base metal sufficiently, especially cast iron. But don't overheat small areas. This may cause the braze filler to fume and boil.

Metals That Can Be Brazed

Metallurgy and metal joining methods are changing as fast as our computers and electronic things are changing. Be constantly aware of new

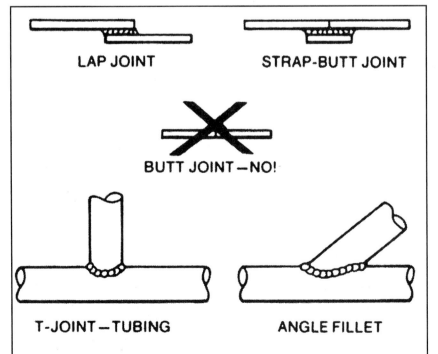

Best and worst joints for brazing, silver brazing or soldering: Never braze butt joints.

materials and new joining methods. Read Chapter 15 for new methods and materials for joining metals.

A wide variety of metals can be joined by brazing. These include stainless steel, cast iron, brass, copper, bronze, aluminum (with aluminum brazing rod), mild steel, cold-rolled steel, chrome-plated metal, cast metal, galvanized steel and other zinc-coated steels.

Dissimilar metals can be joined, such as copper to steel, or copper to brass. This usually is not possible in fusion welding.

Gas Brazing Procedure

You must never breathe brazing vapors. Use ventilation as necessary. When galvanized or zinc-coated metals are heated with a welding torch, fumes are given off that are extremely dangerous if inhaled. These fumes appear white, similar to cigarette smoke. It's safest to avoid welding, cutting, brazing or soldering galvanized or zinc-coated metals. Instead, let an experienced welder do it for you. But, if you feel you must weld galvanized pipe or sheet metal, follow these precautions:

•Weld galvanized metal only outdoors in open air so fumes will not concentrate.
•Or use a commercial air extractor to suck the fumes into a filter and away from humans and animals.
•Wear a high-quality breathing respirator while welding galvanized metal.

When practicing brazing, you'll be doing similar, but fewer, operations than you did when practicing gas-welding steel. Brazing involves three operations instead of five. You won't puddle brass or do butt welds. Instead, you'll concentrate on lap welding.

You'll need several 2" x 5" pieces of 0.030" thick mild-steel scrap. Your three projects are:

•Running a bead with brazing rod.
•Lap-brazing two pieces.
•T-brazing two pieces.

Required Equipment:
•Oxyacetylene welding outfit.
•1/16" brazing rod (36" long, cut in half).
•Powdered brazing flux or flux-coated rod.
•Bucket of water for removing flux from braze bead.

Common Mistake—Many inexperienced braze welders overheat the base metal. After learning to fusion-weld mild steel at 2,700°F (1,482°C), they must learn to braze at 1,050–1,075°F (566–580°C). Remember that all metals start to vaporize at their boiling point. Brass brazing rod will boil, vaporize and generally ruin your project if heated to the melting point of steel!

For best results when brazing mild steel, heat the base metal to about bloodred to dark cherry red—no hotter! To do this with a 6,300°F (3,482°C) gas-welding torch, hold it farther away from the metal than if you were fusion-welding. You may think that a smaller torch tip will give less heat—true. But, for brazing, you should use a large soft flame rather than a small hot or harsh flame.

To get a soft or quiet flame, use a medium tip with very low gas pressures. You'll hear the difference between the soft flame compared to a neutral flame for welding steel. You'll like the way the soft flame sounds.

Flux the Rod—Open the can of brazing flux. Using a 0.040" tip, set both oxygen and acetylene regulators to 2–3 psi. Before you light your torch, double-check that you have on the proper apparel: long-sleeved shirt, long pants, gloves and welding goggles.

Light the torch and adjust for a soft flame; not loud and hot. Brush, or bathe, the working end of the brass rod with the flame to get it warm—not molten. Quickly dip the rod into the powdered flux. When you pull out the rod, flux should have adhered to about 2" of the hot end of the rod, completely covering it. Use plenty of flux. Don't worry about using too much. A can of flux goes a long way. One can will last me years, even when I do a lot of brazing.

Precoated Rods—You can buy brazing rod precoated with flux, eliminating the inconvenience of repeatedly dipping the rod into the flux. Sometimes, even an old pro like me buys some.

The problem with flux-coated rods is that they must be protected from moisture and rough handling. The flux coating can break and flake off. If this happens, there won't be enough flux on the rod to do a good braze job. I buy just enough rod to last 30 days or less, so the rods are always fresh.

Run a Braze Bead—Set your piece of scrap metal on firebricks. If you haven't already done so, light the torch, adjust it to a soft flame and hold the tip 2–3" from the workpiece metal until a 1"-or-so round, bloodred spot develops. Now, just touch the hot spot with the brazing rod. The filler should melt and flow onto the steel. Continue this heating and touching process until the rod needs more flux. Dip the rod back into the flux and keep going.

Brazing Joints

Lap Joint—Now you're going to braze a lap joint and see capillary action at work. Make sure your two pieces of scrap steel are clean and rust-free. They should be flat so the edges will not bow up and look bad afterward. See the accompanying drawing for how to fit the two pieces. Remember to support the workpiece off the table so it doesn't soak up the heat.

Light the torch as before and coat the rod with flux. Play the torch along one end of the seam to heat both pieces to dull cherry red. When you're sure that both pieces are the same color and temperature, touch the edge of the seam with the flux-coated brazing rod. Watch the molten filler flow into the seam! That's caused by capillary action. Continue along the seam until you have it filled end-to-end with filler.

Shut off the torch and let the metal cool for about 3–4 minutes. Pick up the metal with your pliers and dip it into a bucket of water to cool and soften the flux. The flux turns into a glass-like substance after cooling from its molten state. Although you can chip it off, it's easier to let water soften it. Then, it's an easy job to remove the flux with a wire brush.

If you succeeded in getting the filler to flow completely into the seam, the joint should be capable of lifting a car!

T-Joint—After you've mastered brazing the lap joint, practice brazing a T-Joint. Support the pieces so heat isn't absorbed into the work table. Hold the T in place with a "mechanical finger"—page 139. Tack-braze it in place. A tack should go at each end of the T.

As you start brazing the seam, remember to control the temperature of the pieces. Remember gas-welding a T-joint? You cannot heat just the bottom metal piece and expect the brass to flow onto the vertical piece. Manipulate the torch so both pieces are heated equally—dark cherry red. Remember, if you overheat the steel, the brazing rod will fume or boil! After you've finished the seam, shut off the torch.

The seam must be filled with filler to be strong. Adhered surface area must be max-imized. The drawings nearby show the best seams for brass and silver brazing.

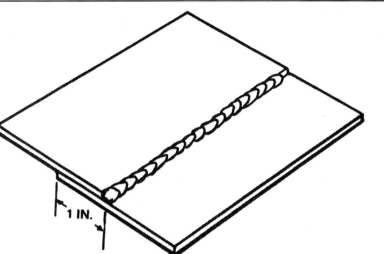

Overlap braze lap joint about 1". Heat joint evenly until both pieces are bloodred. Apply brazing rod and watch it flow into joint.

Several different kinds of flux are required for the many and varied kinds of brazing and soldering processes. Here are some of the types of flux I keep in my corrosives cabinet. Yes, flux is almost always corrosive, but it can be easily neutralized with warm water.

Brazing Aluminum

Brazing aluminum is similar to brazing steel or other materials. It also has the same problem as gas-welding aluminum: The base-metal color doesn't change as it's heated.

The clue to judging correct temperature for brazing aluminum is to watch the flux that's on the base metal. When it starts to melt and flow, the base metal is ready to be brazed.

Use a slightly carburizing flame to reduce aluminum oxidation.

Not only must aluminum-brazing rod have flux on it, you must also apply flux to both sides of the weld joint. I like to use a small metal-handled acid brush to apply the liquid flux, but you could even paint it on with a thick-bristled paint brush. Don't use a brush with plastic bristles—the bristles will melt. Acid brushes are relatively inexpensive—so inexpensive that you could throw them away after one use. For this reason, I keep a dozen or so brushes around at all times.

Several tiny rust pinholes are just brazed shut in this photo.

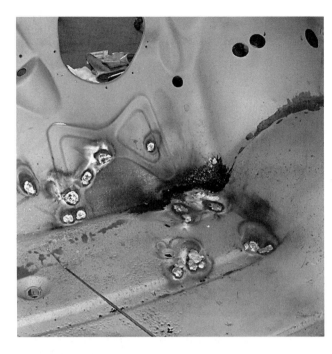

Brazing Copper, Cast Iron & Other Metals

By the time you're ready to braze other metals, you should be able to determine when the base-metal temperature is right for brazing. Copper turns red, stainless steel blue, and cast iron yellow when they've reached the right temperature. But, as with brazing aluminum, the best way to judge when the proper temperature has been reached is to watch the flux. When it melts, the base metal is ready to accept the filler metal.

Silver Brazing

You can silver-braze just about any metal that can be brass-brazed. Silver wets the metal better than brass. It also sticks to some metals where brass will not, such as carbide tool steel. This very hard steel is used for tipping saw blades and other cutting tools. Silver braze also gives a superior appearance to some projects such as costume jewelry.

Because silver is one of the metals in silver-brazing rod, it costs much more than brass filler. A silver-bearing alloy of low-tensile strength—about 20,000 psi—can be used to join dissimilar metals such as aluminum, steel, copper, stainless steel and monel. This particular low tensile-strength silver alloy melts at low temperature—about 500°F (260°C).

Silver-Brazing Procedure—The first thing to do is thoroughly clean the joint surfaces. The joint clearance should be 0.002"–0.006". If you can't judge this spacing by eye, use a feeler gauge to set the spacing at exactly 0.004". After setting up parts

with a feeler gauge a few times, you should be able to judge the 0.002" to 0.006" gap by eye with ease.

Paint the joint area with flux thinned with water or alcohol. Coat both sides. Use a slightly carburizing flame and heat a broad area, keeping the torch in motion. When the flux turns clear and starts to run, add enough silver alloy to completely fill the joint. When finished, shut off the torch and allow the joint to cool for 3–4 minutes. Remove flux with hot water.

SOLDERING

Soldering is another welding operation done below 800°F (427°C). Lead soldering and silver soldering are similar processes. The major difference is that lead soldering is more akin to brazing because of the lower temperatures used. Brazing, lead soldering and silver soldering all require the use of heat, flux and capillary action.

Metals that are easily soldered are platinum, gold, copper, silver, cadmium plate and tin. Less easy to solder are nickel plate, brass and bronze. Metals that are more difficult to solder because they don't wet easily are mild steel, galvanized plate, and aluminum alloys 1100, 3003, 5005, 6061 and 7072.

Electric Soldering

Most people are familiar with soldering copper wire with an electric soldering iron or gun. I concentrate on flame soldering. The reason for this is that electric soldering generally is restricted to

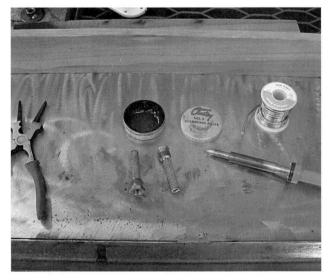

In this picture, we are soldering a couple of stainless steel wire screen strainers to brass pipe fittings to be used in an airplane fuel tank system, by using an electric soldering iron, soldering paste to clean the parts for soldering, and some 50/50 solder that is also flux cored.

copper wiring and similar parts that have a small amount of mass. Although it's a low-temperature welding process, a conventional electric soldering iron, simply won't heat much more mass than a thumbtack to a temperature sufficient for soldering! The temperatures involved are the same, but the quantity of heat is different.

Soldering Procedure

Although welding goggles are not required to solder because intense light is not generated, you should wear safety glasses to prevent eye injury in case the solder pops or splatters. Once the goggles are on and the torch is in hand, follow these steps to flame solder:

•Clean the base metal with a Scotchbrite abrasive pad, steel wool or emery paper. Remove all oil, grease, paint and anything not part of the base metal. Otherwise, the solder will not adhere to the metal.
•Apply flux to the base metal. Choose the correct flux. Use an acid brush to apply it to all surfaces you intend to solder.
•Heat the base metal to soldering temperature by playing a soft flame over the base metal until the flux melts and starts to run. Then touch the metal with solder. If heated correctly, it will flow into the joint by capillary action.
•Apply solder until the joint is filled. Apply heat as needed.
•If necessary, remove flux from the joint. Use a wire brush or water for this.

Sometimes you have to un-solder some steel as John is about to do here, by using an oxy-acetylene torch to melt some old body solder and a wet shop towel to wipe the old, melted solder out of a body seam before welding in a new piece of steel repair panel.

In this photo, I'm driving my SCCA class D Sports Racing car that I built by brazing the 1" diameter steel tube frame, the suspension and most of the other steel parts of the car. It is powered by an 850 cc SAAB three-cylinder engine. Photo Jim Pittman

Chapter 11
ARC WELDING

Small arc welding jobs can be safely done behind a small welding shield, as is being used here. You are responsible for eye protection when you are welding.

Before reading this chapter, you should read Chapter 8 on *Gas Welding and Heat Forming*. As mentioned throughout this book, gas welding should be mastered, or at least understood, before you attempt any other welding techniques.

The first type of electric welding to be invented was the arc welder, and it continues to be a very useful welding tool. The selection of filler metal (welding rods) for the arc welder is extensive, and the rods are very easy to carry and to use. Small fabrication shops and the largest ship builders still find the arc welder to be a very handy fabricating tool.

The best way to become really good at arc welding is to read these instructions, then go practice, then read some more, and then practice, practice, practice. So, let's do it!

ARC WELDING BASICS

This section describes how to set up the arc welder you need and can afford. I say "afford" because arc-welder prices range from $99 to $50,000. You probably don't want or need the cheapest or most expensive welder.

Just as with oxyacetylene equipment, you must determine your welding needs, then choose the arc welder that meets those needs. Remember, every welding project in Chapter 16 can be made with arc welders costing less than $300.

For most of you, a large-capacity arc welder is unnecessary. Even commercial welding shops often adjust their largest-capacity arc welders to about one-third of their capacity. For example, a 400-amp machine usually will be set at 100 to 125 amps.

Duty Cycle

More important than a welding machine's amperage is its duty cycle—the percentage of time the welder can be used at its rated output before it must rest. This rest allows the machine to cool before resuming the weld. Exceed the duty cycle and the machine will gradually develop less than its rated output. A typical duty-cycle rating is 60% at 200 amps. This means the machine can be used for 6 out of 10 minutes while set at 200 amps. Duty cycle goes down when amperage is increased; it increases when amperage is reduced. A cost-efficient choice is a welder with a 90% or 100% duty cycle at 100–125 amps.

When considering duty cycle, keep in mind that you won't be able to weld 100% of the time regardless of what the welder is capable of. You must stop to change rods, take a rest, change positions, reset the welder, or stop to weld another part or weld seam.

Installing a 220-Volt or 440-Volt Welder

You don't have to pay much for a small AC 225-amp buzz-box welder. However, if your workshop doesn't have an outlet to plug it into, it will take you a day or two and more than the cost of the welder to have an electrician wire it in. And no, you cannot plug a 220-volt welder directly into an outlet meant for a 220V electric clothes dryer without adapters (see the photo nearby). Even though the female plug will likely be 220V-50 amp, the plug style will likely be different. You can either change the wall outlet, or you can buy 2 plugs, one male that fits into the wall plug, and one female plug that your welder will plug into. Then you buy

You can make up adapters like this to be able to plug your 220-volt welder into 220-volt dryer or stove receptacles that are different than your 50 amp 220-volt welder male plugs. We used 10-3 S cord (8 gauge, 3 conductor flexible cord) to make the adapters.

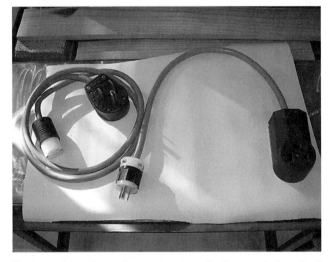

You may have to make your own adapter cords to adapt your new 110 volt–220 volt automatic switching welding and plasma cutting machines. The cord on the left plugs into 220-volt power and makes a combination plasma cutter work with 220 volts or 110 volts. The cord on the right converts a 110-volt–220-volt welder to weld on 110 volts or 220 volts, all automatically. But you have to make these adapter cords to make the equipment work.

20 ft. or so of 8-3 S cord and make up a handy extension cord and 220V adapter. Problem solved!

If you buy a 440-volt welder, it's possible that you'll have to rewire the building's main service box to obtain the power needed for your welder.

Now that you're aware of the possible electrical problems, don't be bashful about buying an arc welder. Once you overcome any wiring problems, final setup is easy.

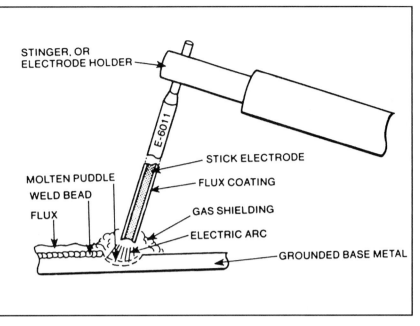

Arc welding generates high-temperature, gas-shielded metal spray to create molten puddle on base metal. Molten puddle solidifies as rod is moved along weld joint, leaving slag-coated weld bead.

Arc Welding Rod

Refer to Chapter 4 for specifics on which fillers to practice with. Pick two or three kinds of welding rod. I suggest starting out with 5 lbs. of E-6011, E-6013 and E-7018.

Storing Arc-Welding Rod—All coated electrodes should be stored in a dry, warm atmosphere. I've seen it stored the following ways:
•In plastic pouches with the ends taped airtight and a couple of packages of desiccant crystals inside. The desiccant absorbs moisture. Although this method keeps the rod dry, the pouches are cumbersome to use when welding.
•In a tube with a tight-sealing lid and desiccant inside.
•In a 5-gallon metal can, or rod oven, with a seal on the lid and a 60-watt light bulb inside. The light burns 24 hours a day. This method works OK, but keeping the 60-watt bulb lit continuously gets expensive.
•In a commercially built rod oven. Because of the expense, this method is practical only for welding shops using over 50 lbs. of rod a month.

When doing certified welding with E-7018 low hydrogen welding rod, dry rod is very important. Weld quality is so important at the nuclear powerplants that if the rod is exposed to the atmosphere more than eight hours, it must be thrown away. Although E-7018 rod can be reheated to drive out moisture absorbed from the air, it is

USEFUL EQUIPMENT

Arc-welder cart. Every welding shop I've worked in had the arc welder on wheels—on a cart or even a trailer. The reason for this is basic. It isn't possible to bring all welding jobs to the machine. Instead, you'll have to take the machine to the job now and then.

An arc welder should be supplemented with a gas-welding rig, including a cutting torch. Otherwise, you would have to do all your cutting with a hacksaw, metal shear or other metalworking tool. To give you an idea of how useful one can be, I use my gas-welding outfit 10 times more often than my arc welder. Here's a list of useful equipment:

•Hand grinder for dressing welds and beveling edges of metal plate prior to welding. A disc sander fitted with a stone or 4" grinder works well for this.

•Bench grinder. This type of grinder is handy for grinding small parts. Either mount the grinder on your workbench, fabricate a stand for your grinder or buy one.

•Chipping hammer. This special hammer is necessary to remove slag from arc welds.

•Several wire brushes. Use a wire brush to clean off the slag after chipping.

•C-clamps. These are like having several extra sets of hands. C-clamps are a must for holding parts together or in position for welding.

•Marking Tools. Turn to Chapter 5 for information about these tools. Regardless of the type of welding or cutting you'll be doing, you'll need to accurately indicate where cuts and welds are to be made.

•Set of hand tools for disassembly and assembly work.

•Spare welder's helmet and lens for a helper to use or for a friend to watch with.

•First-aid kit with burn ointment. You'll need a first-aid kit eventually, even though you are super careful. It's almost impossible to work around hot metal without getting burned now and then. Eye burn fluid is also recommended.

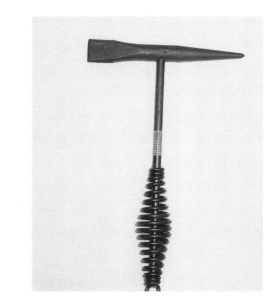

A chipping hammer like this one is a requirement for removing slag from arc welds.

Practice

If you've been collecting pieces of scrap metal, you should have some pieces of 3/16" or 1/8" mild steel. Use a cutting torch to make this 2" x 5" pieces on which to practice. This is a good size—large enough to work with, yet small enough to conserve your scrap pile.

Basic Practice Steps—As you did when learning to gas weld, learn the four basic types of welds before you begin a project. These are:
• Running a bead
• Butt weld
• T-weld
• Lap weld

For practice welding, you need a table with 1/4" to 1/2" thick, 2" x 3" steel top. In a pinch, you could simply lay a steel plate across two wooden stools or sawhorses. Later on, you could build an arc welding and cutting table. See page 137 for plans. Actually, I strongly recommend that your first welding project be this cutting and welding table. It can last you a lifetime, and it will streamline all your future welding projects.

Basic Arc Welding Principles

Before you strike the first arc, you should know what happens at the electrode tip. A 6,000–10,000°F (3,320–5,540°C) temperature is generated by an electric arc between the electrode tip and the workpiece. The flux coating on the welding rod is heated to a gas and liquid. This shields the molten puddle from the atmosphere; thus the name shielded metal-arc welding (SMAW).

never as easy to weld with as it was when fresh and dry.

Of course, all arc-welding rod should be kept dry. I've even seen E-6011 rod fail to maintain an arc because it was improperly stored for several weeks and, as a result, absorbed considerable moisture.

ARC WELDING TECHNIQUES

In this section, I describe how to use AC welding machines. If you want to use a DC machine to practice, read the section on polarity, page 94.

The shield prevents the molten puddle from chemically reacting with or being contaminated by atmospheric gases, which can cause hydrogen embrittlement, porosity and other bad effects.

As the weld puddle solidifies, the flux also solidifies, forming a coating on the weld bead and protecting it from the atmosphere as it cools. This resolidified flux—slag—is glass-like, and can then be chipped off to reveal the weld bead.

In the drawing nearby, you can see the arc welding process with stick electrode. The arc-welding rod actually sprays molten metal into the molten puddle on the base metal.

Remember: Where you point an arc-welding rod is where the weld metal goes! The heat and sprayed metal come off the end of the rod like a spray gun! Point the rod where you want the weld bead!

Striking an Arc

I teach my students how to strike an arc, run a bead, and actually weld something practical in less than two hours! For a complete novice, it only takes five minutes to learn to strike and maintain an arc!

Ready the Welder—The first thing to do when getting ready to arc weld is to ground the workpiece. You can't start an arc without a ground. Either connect the ground clamp directly to the work or to the metal table you'll be welding on. If you don't connect the ground clamp to a suitable ground, you could become the ground!

Once you've grounded the work, adjust the machine to 130 amps. Although this is a hot setting, you'll have an easier time learning how to strike and maintain an arc. Once you've learned to do these two things, you can readjust the machine to a lower amperage setting. Again, to make things easier, you'll need about five E-6011, 1/8"

electrodes for practice.

The welding machine is now ready to be turned on. Make sure the welder is. Regardless, don't do it while the working end of the welder—electrode holder, or stinger—is laying on the grounded table or workpiece. You may see some premature arcing.

Ready the Welder—Don't turn on the welding machine until you've prepared yourself. You must be wearing the correct welding apparel: long-sleeve shirt, cuffless pants, high-top shoes, gloves and a welding hood. A leather apron is not necessary, but a good idea.

Turn on the Machine—Turn the machine to ON. With the electrode in hand—your right hand if you're right-handed, and vice versa if you're a lefty—squeeze it to open the jaws and insert the bare end of an electrode. Usually, there are grooves in an electrode-holder's jaws—for holding the rod at 90° to the holder 45° forward, 45° backward and inline with the holder. For now, position the rod so it's 90° to the holder.

Strike an Arc—Welders compare striking a welding arc to striking a match. However, a freshly lit match is immediately moved away from the striking surface. Not so with an arc welder. Once the arc is struck, you must keep the electrode tip near the work to maintain the arc. Instead of a match, the arc welder can now be compared to a spark plug. If a spark plug gap is excessive, it will not operate. The same holds true with the arc welder. Usually, you should maintain an electrode-tip-to-work gap of 1/8" to 1/4".

With these points in mind, let's get on with the business of arc-welding. With the stinger in both hands and welding hood or tinted shield flipped

A common arc welding bead is this weave pattern. Use it only when making single-pass welds.

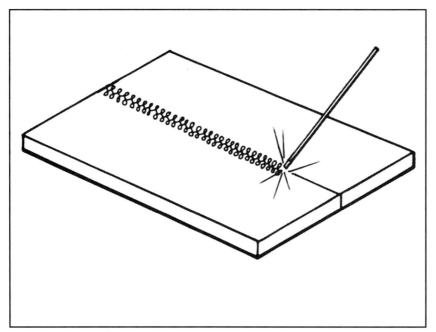

practice running a bead until you're satisfied with its appearance. Use the pictures in this book as examples, or the welds on a car's trailer hitch or trailer. After welding a half-dozen or so acceptable beads on scrap steel, you should be able to try making a butt weld with two pieces of scrap metal.

Butt Welds

Items required for practicing a butt weld include two 2" x 5" steel plates, about 1/4" thick, and two E-6011 or E-6013 welding rods. The plates should be trimmed straight and even so they'll butt without any gaps. Place the two plates on your welding table. Butt them together, then tack-weld each end together.

As the first tack weld cools, the opposite end of the butt-weld seam will open up in a V-shape. Close the seam by supporting the back edge of one plate and tap on the edge of the other one with a small hammer. Once the gap is closed, tack-weld the other end of the seam. If the machine is still set at 130 amps, turn it down to 90 amps at this point in the practice. However, if the rod sticks in subsequent welding, turn up the machine to 130 amps again.

You're now ready to run your first butt-weld bead. Strike an arc at the right end of the seam, if right-handed, and slowly run a weld bead the length of the seam. The weld beam should be centered in the seam. After the weld is completed,

up—depending on the type of hood you have—hold the electrode tip about 1" from your workpiece. You are going to scratch or "tickle" the work with the rod to strike, or start, the arc.

Running a Bead

With a mental picture of where the electrode tip is, nod your head so the helmet will fall down, covering your face and eyes. The next light—let's hope—will be the arc. Like writing with chalk on a blackboard, scratch the work with the electrode to start the arc. You can't just touch the electrode tip to the work to start the arc—the tip must be moving. Once started, keep the small gap just suggested to maintain the arc, and move along slowly. But, chances are you won't get this far on the first few tries.

If the rod sticks to the work—it happens to everyone—swing the stinger from side to side to break it loose. Do this quickly. Otherwise, the rod will get red hot and soft. If this happens and you can't break it loose, squeeze the holder to release the rod. You can then use pliers to work the rod loose. Don't grab the hot electrode with your hands, even if you're wearing gloves!

After learning to strike an arc, you'll have to maintain it by moving along while maintaining the correct gap. To do this, you'll have to move the holder closer to the workpiece as you move along. The rod foreshortens as it melts to create the weld bead.

Once you've learned to strike an arc and maintain it, keep practicing while it's fresh in your mind. The object is to run a good weld bead.

Run the Bead—On a piece of scrap steel,

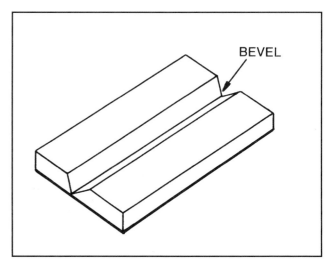

BEVEL

When butt-welding thick material, seam is beveled to obtain maximum penetration.

let it air-cool for about five minutes. You can now pick up the metal with your pliers and stick it in water to complete the cooling.

Check for weld penetration. Look at the backside of the seam. If penetration is good, you should see signs of the weld puddle dropping out of the bottom of the seam or extreme discoloration of the metal. If penetration appears to be insufficient, turn the heat up 15 amps and try again on two fresh strips.

Practice making butt welds until you think they look pretty good. To remove doubt as to the quality of your welds—a good-looking weld isn't necessarily a quality weld—take your best samples to a technical-school welding shop. Ask for the instructor's opinion.

Test Weld—An easy way to test a butt weld is to clamp one side of the weld sample in a large vise, just below the weld seam. Hit the top of the side opposite the weld bead with a hammer. This will bend the metal toward the topside of the weld bead. If the weld is weak due to poor penetration, it will break through the backside of the weld seam.

Common mistakes for beginners are moving the rod too fast, resulting in poor penetration. Slow down! It's better to have too much weld bead than not enough.

For beginners, the best weld bead is obtained by moving the arc-welding rod at the same speed as the second hand on a wrist watch—about 3" per minute. As your skill improves, adjust travel speed to maintain a molten puddle.

T-Welds

Doing a T-weld with an arc welder requires that you manipulate the rod to avoid undercutting the vertical piece and to get good penetration in both pieces of metal. The tendency is to burn through the vertical piece and get insufficient penetration on the horizontal piece. This is the same problem encountered when doing a T-weld with oxyacetylene. The difference between the two types of welds is that you must now manipulate an electrode rather than a gas-welding tip.

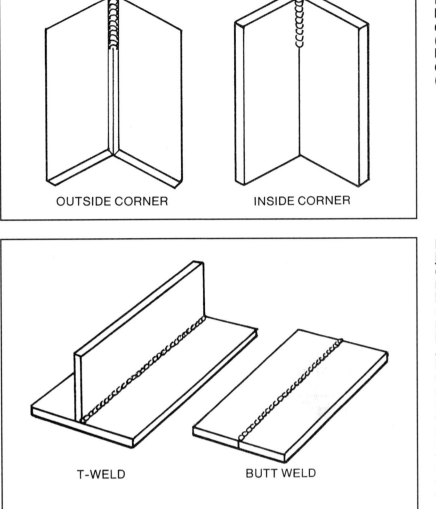

OUTSIDE CORNER   INSIDE CORNER

Different welds take different heat. Outside-corner weld (left) takes less heat than inside-corner weld (right).

T-WELD   BUTT WELD

Different weld-joint configurations require different heat. T-weld (left) requires more heat than butt weld (right). Also, majority of heat must be directed toward piece with most mass. For example, majority of heat must be directed toward horizontal piece when doing T-weld.

To practice doing a T-weld, you need two 1/4"-thick steel plates that measure about 2" x 5". Place the two plates on your welding table so they form an upside-down T. Use a mechanical finger to hold the vertical section in place while you tack-weld each end. If the first tack weld causes the vertical plate to raise at the seam, tap it back in place with your hammer. Tack weld the other end of the seam.

You'll need two sticks of E-6011 or E-6013 welding rod. If you're right-handed, strike an arc at the left side of the seam and start the bead. Apply about 70% of the heat to the flat part and 30% to the vertical part. This means you spend 70% of the time pointing the rod at the flat piece and 30% pointing the rod at the vertical piece. You'll have to swing the holder back-and-forth as you go while maintaining the arc gap and welding puddle. Have fun!

If you're left-handed, start at the right end of the seam and strike your arc on the lower piece.

T-weld undercut is caused by a vertical piece being overheated. Base metal then flows onto horizontal piece. Manipulate electrode so horizontal piece is heated most, but aim electrode at vertical piece. Point the rod where you want the heat to go.

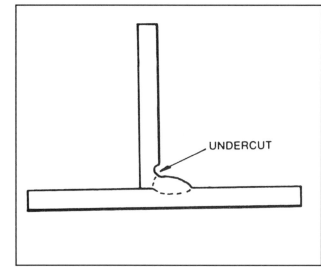

UNDERCUT

Starting on the upper piece would cause slag to be buried under the weld bead on the bottom piece. To time your stinger movement, count out loud, one-two-three, one-two-three, if it helps. You'll be constantly moving your rod-holding hand to accomplish this.

Again, when the weld is finished, let it cool. Chip off the slag and inspect the weld. If the vertical piece is undercut, suspect an excessive arc gap or rate of travel. Do it again on more scrap pieces.

Keep practicing until you're satisfied with the looks of your welds. Once you've mastered the problem of undercutting the vertical piece and can get sufficient penetration in the horizontal piece with the same weld, you can do T-welding.

Lap Welds

A lap weld is the most difficult type of weld to do with an AC arc welder. You'll need two 3/8" thick, 2" x 5" steel plates. Lay them on your welding table so one overlaps the other. Tack-weld each end. E-6013 rod is best for lap welds, but E-6011 is OK in a pinch. As with a T-weld, the trick for doing successful lap welding is to prevent undercutting the top piece.

Again, strike the arc on the lower piece. Weave the rod in the motion shown on page 90. Be sure to pause slightly at the bottom each time you make a swing to prevent undercutting the top piece.

After you've completed running the lap-weld bead, cool the piece and chip off the slag. Inspect the weld for undercutting and penetration. Practice until you are making consistently good lap welds.

OUT-OF-POSITION ARC WELDING

Anything other than welding a seam that lays flat is called out-of-position welding. This applies to all types of welds: butt, T or lap. And, out-of-position welding will separate the men from the boys. Not only are they the most difficult to make, some are more difficult than others. Although I can give you tips on how to do each type of out-of-position weld, the only way to master each is with practice.

Regardless of your experience, always try to weld everything in the flat position, if possible. But many times it's impractical or impossible to reposition an item to make welding easier. So, you weld it where it is. Although there are several tricks when doing each type of out-of-position weld, the number one trick is to point the rod where you want the puddle to go. You're fighting gravity when welding out of position.

I'll never forget some out-of-position welding I once did while on the top rung of a 30-foot ladder. I was hanging upside down by my knees, welding the bottom side of a bracket on a light pole! While making that very uncomfortable weld, I said to myself, "Now this is really out-of-position welding." That was several years ago, and I'm sure the bracket is still holding tight to that light pole.

Overhead Welding

Use E-6011 or E-6013 rod. Aim the rod so it points almost straight up, about 30° from vertical. Also, you should wear tight-fitting clothes. Sparks will fall on you.

After you start the arc, hold the electrode tip closer to the work than the 1/8" gap that's normal for 1/8" rod. I like to push the rod and puddle up against the base metal.

If you see a drip starting, push it back into the fresh weld bead, or if too big, fling it out of the puddle and start over. You may have to stop occasionally to let the weld bead cool. Usually, three or four seconds is enough cooling time. If you stop longer, chip out the slag to prevent slag inclusions in the weld.

Vertical and Horizontal Welding

For practicing vertical and horizontal welds, I recommend a 1/8" E-6013 rod. Other rods can be used. For instance, E-6011 is OK, but E-7018 is more difficult to use. Even though you may be using E-6013 rod, don't think vertical or horizontal welding is easy. This type of welding is the most difficult to perform and get a good weld.

For vertical welding, point the rod about 15–25° upward from horizontal; 30–45° when welding down. If you weld at 90° to the work, the puddle will sag or fall out.

It is particularly important to watch the puddle, not the arc, when doing a vertical or horizontal weld. Otherwise, you won't be able to control the puddle. If it begins to sag, momentarily pull the rod back to lengthen the arc and stop depositing metal. This also cools the puddle slightly. Move back in to continue the weld. This in-and-out movement is indicative of how to do such a weld, particularly vertical up.

Vertical-Up—The puddle movement is up—which is difficult to make because the puddle is below the arc. You can't use the rod to "hold" the puddle. One way to keep the puddle in position is to momentarily interrupt the arc, freezing the puddle so it stays in position. You can also momentarily bury the end of the rod in the puddle, also reducing heat.

Vertical-Down—Puddle movement is down—which makes it easier to control because you can use the rod to push or hold the puddle in position as it tries to fall or drop out. However, penetration on a vertical-down weld is not as good as that for a vertical-up weld. For this reason, vertical-down welds are not permitted when welding certain types of industrial pipe, such as high-pressure steam pipes in power plants.

Finally, don't forget to point the rod where you want the puddle to go. Although you've read it before, here it is again. All it takes is practice, practice, practice.

Stopping & Starting Arc Welds

One thing common to all types of joints or arc-welding positions is the need to stop, then continue the bead. You may have "burned" the electrode to a stub and need a new one, need to change position, or rest yourself or the machine. Whatever the reason, you must get a smooth transition without voids at the end of one bead and the beginning of the next.

To get a good transition between two overlapping weld beads, do this: Stop and allow the weld bead to cool for about two minutes. With your chipping hammer, knock off the slag at the end of the bead. This is particularly important with deeply grooved weld joints. The slag will flow into the groove at the end of the weld, causing an inclusion or void in the weld if the weld bead is continued over it. Therefore, all slag must be removed completely.

Chip out the slag with the pointed end of a chipping hammer. Finish up by wire-brushing the end of the weld bead. When the seam is perfectly clean, continue welding. Overlap the welds by striking the arc in the small crater at the end of the weld bead and immediately continue running the bead once a puddle forms. If done right, you should have difficulty seeing the transition between the two beads.

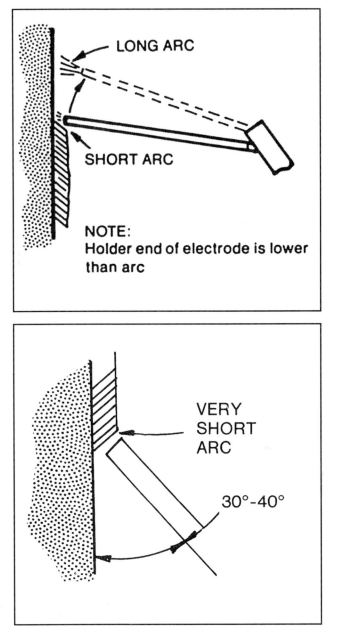

LONG ARC

SHORT ARC

NOTE:
Holder end of electrode is lower than arc

VERY SHORT ARC

30°-40°

Keep electrode horizontal or pointed slightly upward when welding vertically up. As molten metal begins to deposit, move electrode tip 0.5" to 0.75" upward to allow puddle to solidify. Bring tip back to deposit metal. Continue this movement while you only watch the molten puddle.

Maintain 30–45° electrode-to-workpiece angle when welding vertically down. Move fast, otherwise slag will catch up with arc.

ARC WELDING SHEET METAL

Welding sheet metal with an arc welder is difficult to learn because sheet metal is thin and easy to burn through. Here are some tricks to make it easier:
• Weld at low-amperage settings. Try 60–75 amps with 1/8" rod or 40–60 amps with 3/32" rod.
• Hold a very close arc. This keeps excess heat down long enough for the puddle to stick to the base metal.
• Stitch-weld or spot-weld, then tack and fill the gaps. This prevents local heat buildup and burning holes in the metal.
• Use lap welds, if possible. This thickens the metal, creating more heat sink—mass to absorb heat.
• If all else fails, use copper strips of heat-sink

This Lincoln Electric Square Wave TIG 175 Pro Welder can both arc weld and TIG weld all metals that are weldable. It weighs around 200 pounds. Note the DIN connector on the lower right side of this picture It has been superseded by a new model called the TIG 185 that has even more features for small shop use.

compound to back up the weld seam. The weld will not stick to copper; you can remove the strips after the weld has cooled.

ARC WELDING 4130 STEEL

Often referred to as chrome moly, 4130 steel welds similar to mild steel. However, the resulting weld is more prone to cracking after it cools because of 4130's "graininess." Here are some tips for AC arc welding 4130 steel:

•The larger the piece, the more important it is to preheat before welding. Always try to weld 4130 steel in 70°F (21°C) or higher temperatures, and preheat to 200–300°F (93–149°C) in the expected heat-affected zone. Use a temperature-indicating crayon or paint to monitor preheat temperature.
•Preheat with an oxyacetylene torch—rosebud tip if it's a large part—or heat the part in a kitchen oven if it fits. For huge parts such as nuclear power-plant reactors, an electric blanket is used for preheating.
•Always use E-7018 rod for welding 4130 steel, or the recommended rod.
•Make sure the base metal is clean and free of rust, paint and grease. Otherwise, you'll end up with a defective weld.
•Bevel the weld joints to get maximum weld penetration.
•Before taking a chance on ruining an expensive piece of 4310 tubing or whatever, practice on scrap.

POLARITY BASICS

When welding with alternating current (AC), you don't have to set polarity—direction of current flow. In the U.S., AC constantly switches back and forth, 120 times a second, between positive and negative. A complete cycle occurs 60 times a second. In other countries, it may be 50 or 90 times per second. In direct-current (DC) welding, polarity makes a big difference. The above illustrations dramatize the effect of polarity in DC welding.

Almost all DC welding is done with reverse polarity—electrode is positive—because the welding rod gets hotter than the workpiece! Reverse polarity provides a steadier arc, and electrode-to-work metal transfer is smoother than with straight polarity—electrode is negative. It is easier to weld with a shorter arc and low amperage. Therefore, DC is better for making out-of-position welds. Some electrodes can be used with reverse or straight polarity; these are called AC/DC electrodes. And, there are some that can be used only with straight polarity.

The chart nearby gives recommended polarity settings for various metals. Make a copy and keep it near your welding machine so you can refer to it.

Welding 4130 steel is discussed in more detail in Chapter 8, because gas welding is more suitable in most instances. For example, gas welding is better for building a chrome-moly airplane fuselage or race car suspension parts. A gas torch is easier to manipulate around the small parts and preheating is almost automatic.

ARC WELDING ALUMINUM

Although uncommon, it is possible to arc-weld aluminum plate, aluminum castings and aluminum sheet with a DC arc welder. The resulting weld bead will look rough compared to arc-welded steel. I've used it for welding large, thick aluminum plates, building up work edges on aluminum pieces and welding 1/4" aluminum for toolboxes, barbecue grilles and shelf brackets.

As with steel, you should preheat 1/4"-and-thicker aluminum to 300–400°F before arc welding. Expect a very bright arc, and a lot of noise and spatter when using aluminum arc-welding rod. The resulting arc-weld bead will be about 50%

weaker than it would be if it was TIG welded.

ARC WELDING STAINLESS STEEL

Although there are no special tricks to arc-welding stainless steel, don't expect beautiful welds. Stainless weld beads are not pretty unless they are completely protected from the atmosphere. The backside of the weld usually will appear black and rough.

Appearance of a stainless-steel weld can be improved by coating the backside of the seam with flux paste. This protects the seam from oxygen in the atmosphere, minimizing crystallization of the weld.

The best welding processes for stainless steel are TIG and wire-feed (MIG). But if MIG or TIG are not available, you can do an acceptable job with an AC arc welder. Select the correct stainless-steel electrode as outlined above. Again, there are no special methods necessary for arc-welding stainless steel. Preheating is not necessary.

Do the same as you would when welding any material for the first time. Practice on scrap stainless before you try welding an actual part. If you don't have any stainless-steel scraps to practice on, you can use mild steel or 4130 steel scraps with a stainless-steel rod.

If you were wondering, yes, it is possible to arc-weld mild steel, 4130 steel and stainless steel together in one assembly.

ARC-WELDING CAST IRON

You may encounter the need to arc-weld cast iron. Usually, this involves the repair of a cast-iron machine base, farm equipment, transmission case or engine part. The last time I welded cast iron was when a local high school asked me to build a chariot for a football game halftime show. The students wanted it to look like Roman warriors coming into the arena. I was asked to weld late-model car axles to some cast-iron spoke wheels from a piece of old farm equipment.

I used NIROD—nickel-based welding rod—for doing this job. See Chapter 4 for other recommendations on cast iron rod.

It's hard to weld cast iron without it cracking. The reason is its rigidity. When one small area is heated, causing it to expand, the unheated area resists. Unfortunately, the cooler area loses the battle because cast iron is much stronger in

ARC-WELDING POLARITY SETTINGS

Metal	Polarity (AC/DC) in order of preference	Recommended Electrode
Stainless Steel	DC Reverse (Positive)	E-308-15, E-310-15
	AC	E-308-16, E-347-16
Bronze	DC Reverse (Positive)	E-CuSn-C
Aluminum	DC Reverse (Positive)	AL-43
Cast Iron	DC Reverse (Positive)	ESt
High-Tensile Steel	DC Reverse (Positive	E-7010-A1, E-8018-C3
	AC	E-7027-A1, E-8018-C1
Mild Steel	AC	E-6011, E-7014, E-7018
	DC Reverse (Positive)	E-6010, 5P, E-7018

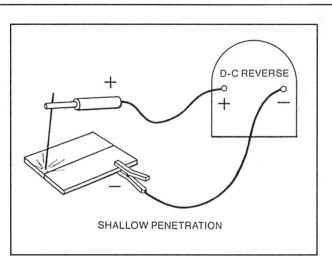

SHALLOW PENETRATION

DC welder set to (+), REVERSE, or POSITIVE polarity, has this circuitry. Welding rod gets hotter when machine is adjusted this way.

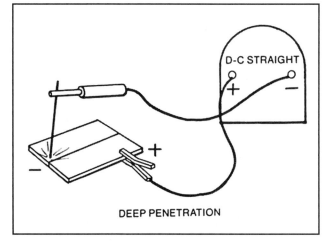

DEEP PENETRATION

DC welder set to (—), STRAIGHT, or NEGATIVE polarity, has this circuitry. Workpiece gets hotter when machine is adjusted this way.

compression than in tension. Thus, the cooler area—in tension—cracks. This is why it is extremely important to thoroughly preheat cast iron before welding it.

As you may suspect, welding cast iron requires a lot of patience. Start by heating the entire casting to

For smaller shops, all you may need are these 2 units from Lincoln Electric Company. On the left is the great new Invertec V205T AC/DC arc and TIG Welder that weighs only 33 lbs. It will automatically switch from 220 volts to 110 volts and can stick weld just as good as any dedicated stick welder. It was used to build the trailer in the first photo in this book. It can also TIG weld at as low as 5 amps for very thin aluminum. On the right is the Lincoln Electric Co. Pro-Cut 25 Plasma cutting machine. It will cut any metal, down to very thin .005" aluminum up to 3/8" thick steel. It weighs only 29.5 lbs. and works on either 110 volts or 220 volts.

400–1,200°F (204–649°C) before welding. Here's another application for temperature-indicating crayons or paint. And, only weld while the casting is hot. This is easy to do with small castings that fit into your kitchen oven, but large castings require a lot of heat. Because of this, it's standard practice for shops that weld cast iron to place a small natural-gas burner under the casting and heat it while it's being welded.

CAST IRON RODS

In recent years, there have been some new developments in cast iron arc welding rod. A relatively unknown company has developed a new cast iron welding rod that really works well and is easier to use than almost all the previous welding rods that were available when this book was first written. The new cast iron welding rod is Cronatron Cronacast 211, part number CW 1034. Look them up on the Internet at www.cronatronwelding.com.

Another approach to welding cast iron—especially big castings that are impractical to preheat—is to arc-weld them at room temperature, but only 1/2"-long beads at a time, then stop. The 1/2" weld is chipped and allowed to cool for two or three minutes before another 1/2" weld is done. This allows the weld and heated area to relax as the heat is absorbed or dissipated into the casting. Some people recommend hammer-peening each short, fresh weld until it is cool.

Brazing Cast Iron

Often, but not always, the best way to join cast iron is to braze it. And you can't do that with an arc welder, so get out your gas-welding torch. I've brazed a lot of small cast-iron pieces with success. The secret is to start by V-notching the crack or joint completely to the center from both sides of the part, or all the way through from one side. After you've done this, preheat the casting to about 350°F (177°C). The welding torch will help maintain heat as you braze. Read about brazing in Chapter 10.

Chapter 12

MIG WELDING

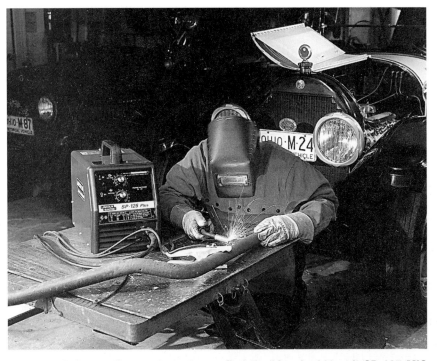

Home workshop antique auto restorers find the Lincoln 110-volt SP-125 MIG welder very useful in making exhaust system repairs. Note that this welder is using a folding picnic table, covered with a sheet of plywood, as a welding table. Courtesy Lincoln Electric Co.

MIG (metal inert gas) or wire-feed welding machines are easy to use. Almost anyone can buy a new MIG welder and be welding just minutes after uncrating it. Because of this fact, MIG welding has become extremely popular in recent years.

When the first edition of this book was printed in 1980, MIG welders were rather simple machines. But electronic inventions and technological advancements revolutionized the welding equipment industry. In 1980, MIG welders were heavy and not very accurate. Today you can buy a very powerful, accurate but rather complicated MIG welder.

So MIG welding machines are available in many sizes and one size does not fit all needs. You really need to know what the welder will be used for during its long and useful life. Small, low-amp MIG welders can never be upgraded to do heavy work. Rather large MIG welders can be tuned and adjusted down to do light welding, as well as somewhat heavy work, however, many of the large units cannot be tuned down and slowed down to do thin-wall, small diameter tubing and autobody sheet metal.

The major advantage of MIG welding is its simplicity and speed. If tested beside any stick welding machine for 1 hour, the MIG machine would be able to weld 4 to 10 times more than the stick-welding machine.

Several years ago I wanted to increase production rates of aircraft seats for a six-passenger, twin-engine airplane. The first 400 ship-sets of seats (2,400 seats) had been welded with TIG, and it was really hard for the welding department to keep up with airplane production. It took one welder 8 hours to build one seat with TIG welding.

So we purchased a high-quality MIG welding machine and had the welders take certification tests with it. After a week, we were ready to build certified aircraft seats. Production of seats instantly went up. One welder could build up to 4 seats in 8 hours with the MIG welder. That is a 400% increase in production.

THE DOWNSIDE

There was one drawback, however; the MIG welds were not as pretty as the old TIG (tungsten inert gas) welds. But the seats all passed Magnaflux inspection, and 9 gs crash testing. However, when we later tried to weld tubular engine mounts with MIG rather than TIG, we found that the corrosion protection oil inside the engine mount tube leaked out at several places because of *cold starts*. A cold start is a lack of penetration where the weld bead appears to cover a seam but is only lapped over until the bead progresses.

MIG welding is notably less accurate than TIG welding and somewhat less accurate than arc welding. At recent race car shows and welding trade shows, builders and buyers were heard commenting about the inherent lack of accuracy of MIG-welded tubular framework as found in race car frames and aircraft frames. The problem is that the weld bead will look good, when in fact 80% or more of the bead is on one

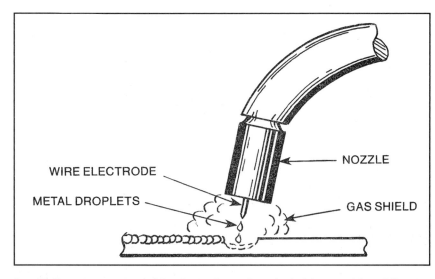

Small-diameter, consumable electrode—wire—is fed into weld puddle at a high rate, or up to 700 in. per minute (ipm). Instead of argon shielding gas typically used with TIG welding, carbon dioxide (CO_2) is preferred for MIG welding. Shielding gas is not required when flux-cored wire is used.

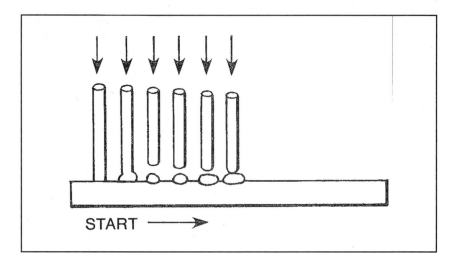

START →

Arc starts at left and weld bead moves to right. Wire continuously advances and melts as it contacts base metal. With machine adjusted correctly, cycle occurs smoothly and quickly, sounding as if it were bacon frying.

member and not on the other member, making this weld only 40% as strong as it should be. We found the same problem when trying to MIG weld the engine mounts on twin-engine General Aviation airplanes.

WHY LESS ACCURACY?

The most common reason is that the MIG gun hides the weld puddle. The solution to that problem is to view the puddle from the side, rather than from behind the MIG gun.

The second reason is the bright light produced by the MIG weld process, combined with the smoke and spatter, make the weld puddle much harder to see than with other types of welding. A partial solution is to install a small, high-intensity headlight on the gun, and to weld with gas rather than flux-cored wire. A third solution is to use

high-quality welding wire rather than the cheapest wire you can find. Read Chapter 4, to learn more about the differences in welding rod and wires.

So, can the accuracy problem be overcome? Yes, it can—read on. But MIG is inherently less accurate than TIG.

BASIC WIRE-FEED OPERATION

Here's how a wire-feed welder operates: The gun is positioned over the weld seam at the same angle you would hold an arc-welding rod. Cup-to-work distance should be about equal to the distance across the cup opening. The gun trigger is then pulled, activating the DC current, positively charged wire electrode—reverse polarity—and gas flow. (Straight polarity is rarely used because the arc is unstable and erratic. Also, penetration is lower.) The wire is simultaneously fed through the gun nozzle and contacts the grounded base metal, causing a short circuit and resulting arc to start. Resistance heating melts both the base metal and the ends of the electrode. The wire then melts back faster than it is being fed to the base metal, momentarily breaking the arc and depositing metal. The arc force flattens the molten metal.

But the wire electrode is still advancing into the puddle, repeatedly arcing and melting off again and again. This on-and-off process occurs about 60 times per second, causing the characteristic buzz you hear. Some people describe a properly adjusted MIG welder as sounding like frying bacon.

MIG welding is called gets its name—metal inert gas—from the types of electrode and shielding used. Unlike TIG welding, a MIG welder uses a consumable metal electrode. This electrode is a continuously fed wire that exits from the center of the welding torch, where a TIG welder tungsten would normally be; thus, the name wire-feed. Typical wire sizes are 0.024", 0.030", 0.035" and 0.045". Up to 1/16" diameter can be used with special equipment. CO_2 shielding gas is used in place of argon because CO_2 is less expensive. Hollow flux-core electrodes are used frequently, eliminating the need for gas shielding, but with lots of smoke and spatter.

Fast & Clean—The major advantage of wire-feed welding is that it's fast. Unlike arc welding or TIG welding, you rarely have to stop for a new welding rod. Also, its weld rate—inches per hour of weld bead—is fast, especially when compared to TIG. Another advantage of the MIG welder is clean welds—much cleaner than those possible from an arc welder, when used with gas shielding.

Modes of MIG Welding

Short-Circuit Transfer—was the original method of MIG welding. It still exists today in the 110-volt machines and in the more basic types of 220-volt machines. See the top drawing on page 98.

Globular Transfer—means that the weld wire only touches the metal when the weld begins. After that, globs of molten wire are expelled into the puddle. Globular transfer occurs especially with higher voltage, CO_2 gas and mild steel electrodes.

Spray Arc Transfer—occurs at higher amps and volts and wire-feed speed, and with argon shielding gas. Higher metal-deposition rates occur, and the arc gives out a higher frequency humming sound. Spray arc transfer is desirable in the flat position.

Pulsed Spray Transfer—is possible when using a special welding machine that is designed to provide optional pulsed arc. Because the pulsed spray operates at two heat pulses, the weld puddle is allowed to freeze slightly between pulses. This feature provides for better control of thin sections, for welding aluminum, and for welding out-of-position with steel wire.

Spot Welding—is possible with a MIG welder by adding a spot-welding timer to the machine. And the welder can also spot weld with short, quick trigger pulls of the MIG gun. But this is not as clean a method as a specific-purpose spot welder. See Chapter 15 for spot welder information.

WIRE-FEED MACHINES

The MIG welder is a simple, compact welding machine. It consists of the welding gun, power supply, wire-drive mechanism and control unit, shielding gas supply and, for some heavy-duty units, a water cooling system.

Gun

In place of the TIG torch or arc-welder stinger is a gun. The typical gun looks like a pistol and directs the filler metal and shielding gas to the weld seam. Service lines running to the gun include an electric-power cable, electrode conduit, and gas hose, if used. Heavy-duty industrial-type guns also have water lines for water-cooling. Otherwise, gun cooling is done with air. Electric power is transferred to the wire electrode via a sliding contact with the copper electrode guide tube in the gun.

The gun nozzle, which is usually interchangeable, determines the gas-shield coverage of the weld puddle. Nozzle-orifice size varies from about 3/8" to 7/8" (10 to 22mm). A larger orifice gives additional shielding, as does a larger TIG torch cup.

So-Cal Airgas makes this Gold Gas™ mix of argon and oxygen for MIG welding. Ask your gas supplier to advise you on the best mix to use for MIG welding.

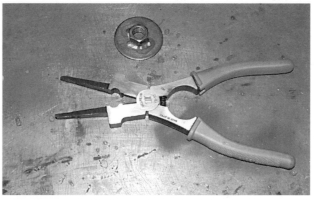

For MIG welding these special pliers are really handy to have. They cut and pull wire, clean out nozzles, remove collets and cups. Ask your welding supply dealer for a pair of these.

Nozzle Dip or Nozzle Spray is used in MIG welding to make the inevitable spatter easier to remove from the gun nozzle. Allan Hancock College.

Power Supply

Almost all wire-feed welders supply DC current. This requires a transformer-rectifier when using an AC power source. Depending on the machine, output can range from 15 to 1200 amps. Required power supplies typically range from 110 to 200/230 volts, or all the way up to 575 volts, depending on machine output. Duty cycle is either 60% or 100%.

Wire-Drive Mechanism & Control Unit

The wire-drive mechanism is relatively simple. It consists of a wire spool and DC motor powered drive rolls—two wheels that run against each other

A crew member from Chip Ganassi Nextel Cup team is welding some new brackets the front clip of this race car with a MIG welding gun. Photo courtesy Lincoln Electric Co.

Notice the welds on this Ducati motorcycle. The weld beads are thick and rounded, more for looks than for strength, although you do want strong frame welds when riding at over 150 mph! MIG welds have a unique look that distinguishes them from TIG or oxyacetylene welds.

It is possible to MIG-weld aluminum and to weld by either gas shielding or flux core shielding as you need to do, but you must change out the liners in the MIG gun to make those changes. The gun liner kit on the left is for the smaller Lincoln MIG welding machines that are to use flux-cored wire. The liner kit on the right and the drive roll and tensioning clamp are for welding aluminum, all with the same Lincoln Electric Co. MIG guns.

with the wire in between. Sometimes, two sets of wheels are used. The drive-roll mechanism pulls the wire off the spool and pushes it through the conduit to the gun at a welder-adjusted rate.

The MIG welder control unit regulates arc starting and stopping, as well as wire-feed, gas flow and, sometimes, water flow rates. It also synchronizes these functions. Usually, there's a jogging feature that feeds the wire to or through the gun while not welding.

Similar to the wire-jogging feature, the control unit also has a shielding-gas purge switch to manually control gas flow. In addition, timers control preweld and postweld gas flow automatically. The purge switch can override the automatic timers. Another timer controls water flow, if water cooling is used. Finally, a wire-feed brake stops the electrode the instant the gun switch

is released. This prevents wire from being fed to the puddle when the arc is interrupted.

Shielding Gas

Except for CO_2 used in place of argon, a MIG welding shielding-gas setup is similar to that used for TIG welding. Not only is CO_2 less expensive than argon, it also has superior heat conductivity.

Regardless of the gas used, constant pressure and flow must be maintained while welding. Also, you must be able to adjust pressure and flow for different applications. Therefore, the gas cylinder must be equipped just as if it were used for TIG welding. Although different gases use different flow meters or flow gauges, the CO_2 must have its pressure and flow regulated.

Types of Wire Feed Welders

When considering a wire-feed welder, remember the old adage: You can't drive a railroad spike with a tack hammer or a tack with a sledgehammer. You must match the machine to the job. If you have both heavy and light-duty welding to do, you need two welding machines, one heavy and one light.

Check out all the MIG machines that are available today. Take your time in making your decision about which MIG machine to buy. There are many options. Several MIG machines can have add-on modules installed so that you can also TIG weld with your MIG power supply.

110-Volt—There are many acceptable 110-volt

wire-feed welders available. Hobart Brothers Company makes a light-duty portable wire-feed welder for sheet metal and body shop work that plugs into 110-volt service. This welder has some limitations, but it's effective at doing what it was designed for—welding 24 gauge (0.0239") sheet metal. It will not weld heavy-gauge stock, such as that used for trailer hitches or farm equipment. Some of the specs for this machine are:

- 100-amp maximum.
- 0.024" and 0.030" steel wire.
- 0.035" and smaller aluminum wire.
- Gas-flow timer allows making stitch welds—tack welds about 1/2" to 1" long—and tack welds. Timer guarantees faster gas shutoff after each weld is completed.
- Wire-feed speed control regulates arc heat.
- Unit is self-contained except for CO_2 shielding gas cylinder. CO_2 is used instead of argon because it's effective and inexpensive. Where high-quality welds are required in 4130 steel, a mixture of 75% CO_2 and 25% argon may be used. This comes premixed in one cylinder from welding supply stores.

200/230 Volt—Several companies make conventional 200/230-volt wire-feed welders. These are the most common machines. Most machines this size can be made portable, but rarely will they plug into a wall socket. Chances are you'll have to change the socket to fit the plug on the machine. They can handle 0.024", 0.030," 0.035" and 0.045" wire.

No CO_2 Gas Required, Your Choice—If you use flux-cored wire—0.045" diameter—no gas shielding is required. It's similar to stick welding, except you don't have to stop for new rod. Flux is on the inside of the electrode, so it doesn't chip and flake off.

When welding any kind of sheet steel, plain or galvanized, this wire gives good weld performance with spatter. It can also weld thicker metals at higher volt and amp settings.

Jason Woods of Alamogordo, NM, leans on the front fender of his friend, Coach Lindley's 1969 Ford Mustang that is being prepared for paint after lots of MIG welding bodywork. The car took a first place trophy in a car show after the job was done.

Ray Ciriza is a laser eye surgery technician for Dr. Schuster in El Paso, TX. He also built this winged super modified race car, using a MIG welder to construct the tubular frame.

575-Volt Heavy-Duty Welder—The minimum these 800-amp machines will weld is 100 amps. They have a 100% duty cycle at 800 amps! You can operate the machine at 800 amps all day, without cool-down, if necessary. If you have some heavy-duty welding to do, this type of machine will handle the job. But it's definitely not for the average home welder! If you think you absolutely must have one, try it first. Remember, you don't need a sledgehammer to drive tacks, and you don't need a 100-amp-minimum welder to build a utility trailer.

MIG-WELDING STEEL & ALUMINUM

Most welding instructors give you about five minutes of discussion about a MIG-welding machine and then tell you to go run a few beads on a piece of steel or aluminum. And that will be the extent of the instruction. That's how easy it is to start wire-feed welding.

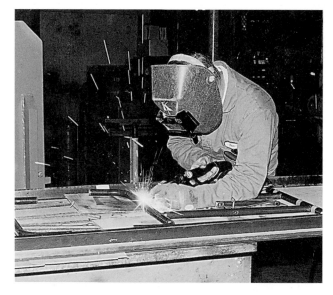

Properly dressed for MIG welding, Ron Chase of Western Welding Co. shows how to repair a folding table with a gas-MIG 220-volt machine. Note that Ron's left hand is guiding the gun for accuracy.

However, there's more to wire-feed welding than pointing the gun at the weld seam and pulling the trigger. Problems can occur if you don't take precautions. The wire may get tangled in the drive mechanism or stick in the collet. The cup on the gun may fill with spatter, making weld quality erratic. Just call it Murphy's Law of MIG welding.

Adjustments

There are also several adjustments that have to be exactly right or the weld will end up looking worse than if it were done with an E-6011 arc-welding rod in the hands of a beginner. Here are some adjustments you'll have to make:

• Wire-feed speed.
• Power—amps.
• CO_2 or argon gas-flow rates.
• Wire size.
• Amps and volts while welding

You'll need a helper to monitor these readings as you weld, and make adjustments as necessary.

Preparation

Before you start welding, attach a small notepad or a sheet of paper to the MIG machine to record settings that work for you. Copy the chart on the next page to help record your settings, or make one similar.

Settings vary from one machine to another. Even though two machines are the same model, they are not the same machine. Some examples of these variables are:

• How fast the wire filler rod is fed at a setting of 1, 2 or 3.

• How hot the wire gets when the volt and amp settings are adjusted to mid-range.
• How much shielding gas is supplied to the weld puddle when adjusted for 20 cfh.

Not only will one machine vary slightly from another, each project will vary in adjustment requirements. For example, if I were describing how to drive my car, I couldn't tell you how far to depress the accelerator to maintain a certain speed. It's the same with MIG welders. Although you can come close on your initial settings, trial and error are required to get them exactly right. So pick a reasonable starting setup. Fine-tune the machine after making a few test beads. Practice on some scrap metal pieces of the same metal you're welding until you can determine the right settings.

As you find the correct settings for each welding situation, record them on your "Wire-Feed Adjustment" chart on the next page. Then, next time you're welding 0.020" steel or whatever, you can set up your machine perfectly by just referring to your chart.

Adjust by Sound—I adjust my welder by listening to the arc. As I mentioned earlier in this chapter, when it's right, the arc sounds almost like bacon frying on a grille over an open fire. Of course, you should also check for proper weld deposit and penetration. Once you're satisfied with the arc sound and weld quality, have someone read the volt and amp gauges while you are welding. Record these numbers in your chart, too.

Wire Cutters—Always keep a pair of small diagonal cutters handy. You'll soon learn why. When you pull the trigger to start the arc and it doesn't, you'll end up with a "pile" of wire. Even though the arc doesn't start, shielding-gas flow and wire feed do. You can't release the trigger fast enough to prevent this, only minimize it. Before you can continue welding, you must cut off the excess wire. Simply snip it off with cutters and continue welding.

Maintenance—Lack of maintenance causes frequent wire-feed welding problems. The cup will get dirty—it must be kept clean. The cup and nozzle must be cleaned of spatter regularly. Special sprays and jellies are made for this purpose. To keep your MIG welder operating trouble-free, perform these simple maintenance operations:
• Keep the cup clean. Normal weld spatter will clog the cup quickly, blocking gas flow and therefore cause the welder to be unprotected from air. To prevent this, you should coat the inside of the cup with an anti-spatter spray or gel such as Nozzle-Kleen or Nozzle-Dip Gel from Weld-Aid Products.

As a less-expensive alternative, many welders use Pam, an aerosol cooking-oil substitute available at supermarkets. If the nozzle does get dirty, clean it. Special ream-like MIG welder nozzle cleaners are available.

• Keep the drive gears and rollers clean. Copper-plated wire will clog the drive rolls in a hurry, so avoid copper-plated wire, if at all possible. Every 2–3 hours of welding time, check the drive rollers for metal filings, and brush them away before continuing. To reduce wire drag and clogging, lubricate the wire. Use a treated-felt applicator to lube steel wire; untreated applicator for aluminum wire.

MIG Welding Stainless Steel

The correct wire for MIG-welding stainless steel depends on its alloy. In most cases, 300-series MIG-welding wire will work with the more common 300-series stainless steel. If the alloy is unknown, try ER-308, a general purpose stainless wire. You can even weld mild steel with ER-308 wire. The weld on mild steel will be much less ductile than the base metal. Some common stainless steel wire numbers are ER-308, ER-309, ER-310, ER311, ER-348, ER-410, ER-420, ER-430 and ER-502.

Each number represents different carbon and alloy content. For instance, the chemical content of ER-308 is carbon 0.08%, chromium 20%, nickel 10%, manganese 2%, silicon 0.50%, phosphorus 0.03%, sulfur 0.03%.

NOTE: Most high-strength steel used in late-model cars is confined to body structures, reinforcements, gussets, brackets and supports. In most cases, the outer panels remain regular mild steel and can still be gas-welded or brazed.

WIRE-FEED ADJUSTMENT

Metal Thickness (in.)	Amps	Volts	Gas Flow	Wire Size	Wire Speed
0.020					
0.030					
0.040					

Make copies of chart for recording MIG-welder setups. You'll then be able to make quick setups when doing similar jobs.

HIGH-STRENGTH STEEL CAR BODIES

To reduce weight for improved fuel economy, many late-model cars use high-strength steel in the body structure. Because the high-strength steel has excellent tensile strength—about 40,000-120,000 psi compared to mild steel's 30,000 psi—panel thickness can be reduced, resulting in considerable weight savings.

But there's a hitch. High-strength steel cannot be gas-welded or brazed. It will harden and crack. Because of its unique grain structure, it must be arc-welded with a low-hydrogen stick electrode such as E-7014, or wire-feed welded. Ford Motor Company and Chrysler Corporation, for example, recommend MIG welding. Here are some of the additional reasons they give for recommending the wire-feed process:

•Welds are made quickly on all types of steel.
•Low current can be used, resulting in less distortion of sheet metal.
•No extensive training is necessary.
•Wire-feed equipment is no bulkier than a set of oxyacetylene cylinders.
•MIG spot welding is more tolerant of gaps and misfits in seams.
•Severe gaps can be spot welded by making several spots atop each other.
•Simple to weld vertically and overhead.
•Metals of different thicknesses can be welded easily with the same-diameter wire.
•Almost all autobody sheet metal can be welded with one wire type.

A FEW MIG-WELDING SAMPLE PROJECTS A 1964 CORVAIR MONZA

1. My wife, Gayle, purchased this 1964 Corvair Monza Convertible a few years ago, not realizing that it had been sold new in Syracuse, NY, and driven in the snow and salty streets there for another several years. It looks straight in this picture, but underneath is a mess of rusty steel.

2. As you could expect, the driver's-side floor was a real mess of rusty metal that had to be replaced. Fortunately, a company called Aftermarket Suppliers makes new sheet metal for Corvairs.

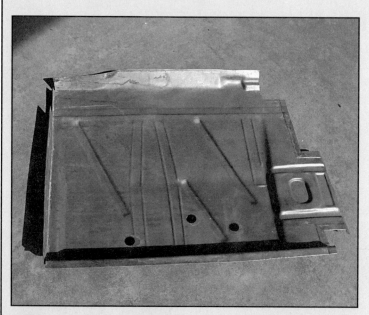

3. This new sheet metal driver's-side floor panel contains more than enough sheet metal to cover the holes in the floor of the Monza convertible. The next step is to cut off the unneeded metal parts.

4. Jason Woods of Tularosa, NM, gets under the Monza to mark the new replacement floor panel where it needs to be trimmed to fit. The work pit in the concrete floor is really handy for jobs like this.

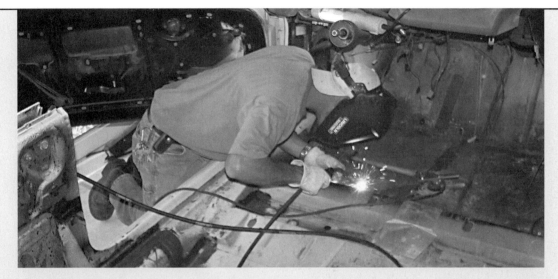

5. Jason begins MIG-welding in the replacement floor panel, keeping the torch cable as free of kinks as he can. The rest of the floor is rather rust-free, thanks to the car being undercoated when it was sold new.

1956 FORD PARKLANE RESTORATION
The following pictures illustrate the type of project a home auto enthusiast can do with a MIG welder and the techniques discussed in this book. The project is a restoration of a 1956 Ford Parklane 2-door station wagon. We'll only show you part of the process. The equipment used was a 175 amp MIG welder.

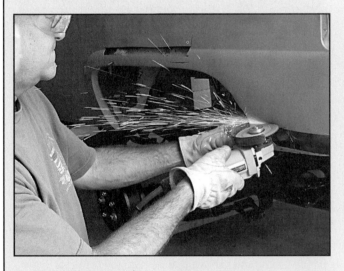

1. John Gilsdorf uses a 4" angle grinder with a cut off wheel installed to cut out a rusty area on his 1956 Ford Parklane station wagon. Note the cut out area above the wheelwell also.

2. John is MIG-welding a wood screw to a dented area so that he can use a special slide hammer dent puller to pull out a dent that he can not get to from the backside.

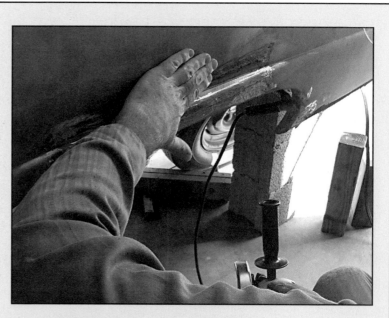

3. The wood screw that was MIG-welded to the crease in the lower body is being pulled out to shape with a special fitting on a slide hammer. A slotted piece of 3/16" steel with a slot in it is MIG-welded to a tube that is then welded to a nut that fits the slide hammer, thus making a suitable tool to pull dents out of blind areas. This saved about $350 in the cost of a special tool that spot welds nails to do the same thing.

4. The old delicate fingers are used to check to see if any more pulling needs to be done before moving on to the next dent and rusted-out areas.

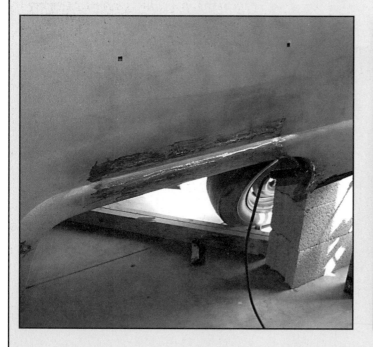

5. In this picture it is hard to tell that there was a rusted-out spot or a crease. But there is a lot more to cut out and weld up before John is ready to paint the body parts.

6. John is finger-testing the surface that he has just repaired. There was an inner fenderwell that kept him from getting behind the area for easy straightening, but it is looking better!

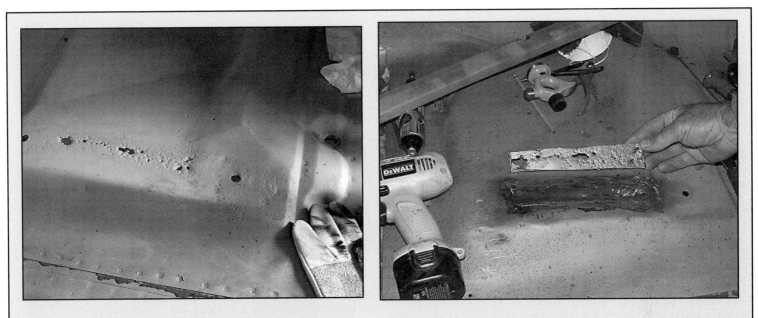

7. This rust area is under the driver's-side seat and it must be repaired by cutting the hole area out and replacing it with a piece from a donor car that John found at a wrecking yard.

8. And here is that same rusty area that has been cut out with a cutting wheel and replaced with a piece from a donor car, by MIG-welding in the replacement piece.

9. John decided that the tiny holes just under the driver's feet were too insignificant to cut out, so it was decided to just laminate this piece from the donor car by glueing it in over the top of the same area of the front floor. It will be twice as thick now.

10. The area under the tail hatch was quite rusted out and it was necessary to take the same sheet metal area out of the donor car to weld in to the area. Thank goodness for donor cars!

11. Sometimes it is better to just use lead to fill dents or even use Bondo to level out the right-rear fender areas. No cutting or welding was done here.

12. It was necessary to fabricate the bracket under the power steering pump to attach it to the water pump. The same MIG welder that was used to weld the sheet metal was used to weld up this same bracket.

This crew member for Chip Ganassi's Nextel Cup racing team is TIG-welding an H-pipe for one of their race cars. The pipe is made from stainless steel. It will make a nice, strong weld if he can "back gas" the pipe to prevent crystallization of the backside of the weld. Courtesy Lincoln Electric Co.

Years ago, TIG (tungsten inert gas) welding, also known by the trademark name Heli-Arc welding, was thought to be magic! It definitely had an aura about it. When I was asked if I would like to try my hand at TIG-welding aluminum, I responded with a big Yes! The next day, I was welding patio screen-door frames for government housing. Nothing to it. It took less than one hour of help from a professional TIG welder and about 5 hours of practice to get the hang of it. Afterward, I took the test for certified aircraft welder—and passed!

Passing the certification test for TIG welder was relatively easy. I was already an accomplished oxyacetylene welder. The point is, you should become proficient at gas-welding before trying TIG. Gas-welding requires the same basic skill—controlling the puddle, moving the torch, dipping the rod, running the bead, etc. And, just like riding a bicycle, these are skills you won't forget as you apply them to TIG and other advanced forms of welding.

CONFUSING TERMS

TIG welding is called several things, which can be confusing. Some people prefer to call it Heli-Arc instead of TIG. Others say it should be called GTAW. The company Linde developed the trade name for TIG welding, Heli-Arc, from the words helium arc welding. Helium is an inert

shielding-gas that envelops the weld puddle, keeping it free from atmospheric contamination. Today, helium has been replaced (largely) by argon, argon/hydrogen or argon/helium mixtures. However, the name Heli-Arc stuck in common usage, even when referring to TIG welders manufactured by other companies. Consequently, most welders have to stop and think about what TIG means—tungsten inert gas or, more specifically, GTAW for gas tungsten-arc weld. It's important for you to understand that these three terms refer to the same welding process, not three different ones.

This section describes various TIG welding setups, from the expensive ones to setups that'll get you by. I start with the very best. As with other welding processes, you should know what you really need before you buy. Because a TIG outfit can be the most expensive of all welders, don't buy one until you have it explained and demonstrated to your satisfaction.

HOW TIG WORKS

TIG is the neatest, most precise and controllable of all handheld welders. You could almost weld a razor blade to a boat anchor or shim stock to a crankshaft.

A small, pointed tungsten electrode—nonconsumable—provides a concentrated high-temperature arc with pinpoint accuracy. You don't have to heat the whole area to start a

In TIG welding, like gas welding, the filler rod is dipped into the puddle while the tungsten electrode is maintaining the puddle. The tungsten must not touch the puddle and the filler rod must not touch the tungsten, or weld contamination will result. When this happens—and it will—stop, clean the weld, clean the tungsten, and start over again.

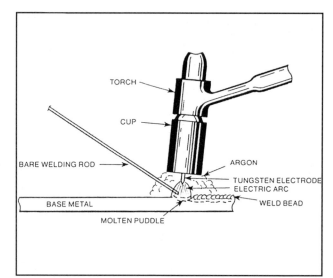

puddle. Once the puddle starts, add filler just as you would with a gas welder.

Because of the TIG welder's high-heat concentration, but reduced heating of the workpiece, it is great for welding aluminum. Aluminum dissipates heat quickly, and the less heat absorbed by the aluminum, the better for the weld. If your workshop has a TIG machine, you could use it for most fusion-welding jobs except for rough ones such as building a race-car trailer. In fact, it can replace the gas-welding torch for all jobs except brazing or soldering.

TIG welding has one major drawback—it's slow. So, for projects that don't require pretty welds but need to be done quickly, use an arc welder or MIG welder.

TIG Components

Torch—Although more complex than a gas welder, the working end of the TIG welder is also called a torch. Instead of a flame, an electric arc is directed at the work to make and maintain the weld puddle. The arc occurs between the tungsten—a high melting point, nonconsumable electrode in the torch—and the workpiece. A collet in the torch clamps the tungsten so it can be adjusted in and out of the torch and retained in place.

Surrounding the tungsten is an open-ended cup that directs this shielding gas to the immediate area of the weld bead. Cups are ceramic because of the intense heat. Speaking of heat, TIG welding torches must be cooled because of the close proximity to the weld puddle. It's not uncommon for a torch to get so hot that it's too uncomfortable to handle.

Torch and cable cooling are done with air or liquid—usually water. Water-cooling is preferred by the serious user. However, an air-cooled torch is suitable for doing small jobs. Air-cooling is not

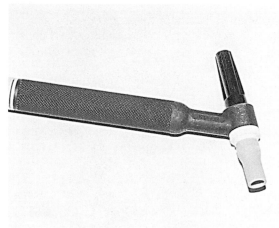

This WP-10 Weldcraft TIG torch is water-cooled and works great for all race car parts, airplane parts, and even ocean-going yacht railings. I like to use a short back cap as often as possible.

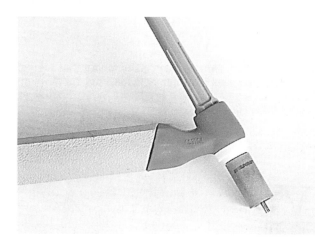

A larger air-cooled torch with a long back cap is useful for field welds on pipe and thicker steel and aluminum materials. Sliding amp control can be added to most torches.

The world's smallest cup, smallest tungsten torch is nice to have in your tool kit when doing those hard-to-reach welds. The cup is heatproof Pyrex glass.

sufficient when welding for long periods or for welding thick material. How do you know when the torch gets too hot? Simple. It will burn your hand. The cable can get hot enough to melt the insulation!

Water, inert gas and electric power must be fed to a water-cooled torch. Consequently, the torch has a cable for electric power and three hoses—one each for gas, water supply, and water return. Water is circulated through the torch and returned to the reservoir or dumped so the torch will receive a continuous supply of cool water.

Cups—The ceramic cup used on a TIG torch directs inert-gas flow over the weld puddle. A larger cup gives more gas coverage and improved weld quality. There are, however, times when a large cup such as a #10 will not fit into a corner or other tight area. Consequently, a smaller cup must be used. Don't go smaller than a #4 cup. Gas coverage will be inadequate.

I use a #10 cup for flat seams, #8 cup for welding 1" diameter tubing, engine mounts and race-car suspensions, and #6 or #4 for tight corners of aluminum air boxes and oil tanks.

Back Caps—These are used to clamp the collet and prevent the opposite end of the tungsten from arcing. Back caps are available in various lengths; short ones are used to weld in tight corners.

COMPLETE TIG SETUPS

If money is no problem, several companies sell complete, first-class TIG-welding outfits. Such outfits will have built-in features to make welding easier. Here are some of these features:

High Frequency: Older Welding Machines

High frequency is provided to start the arc by jumping a spark gap like a spark plug. This is done by superimposing high voltage on the welding circuit. Otherwise, you would have to touch the tungsten to the base metal to start the arc. Touching the base metal is not desirable because the tungsten tip usually breaks off and ends up in the weld puddle. If it breaks off, the result is a contaminated weld bead. A broken tip also shortens the life of the tungsten. Touching the base metal may also contaminate the tungsten.

Although it is possible to weld steel without high frequency, it is required to weld aluminum or magnesium on older non-inverter and square wave machines.

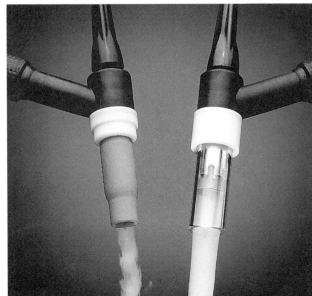

Gas flow, simulated here with smoke, is obviously better with the torch on the right, compared to gas flow on a conventional torch on the left. The glass cup also aids in seeing the TIG weld puddle better. Photo: CK/Conley Kleppen.

Inverter Technology: Newer Welding Machines.

This is the feature that lets the manufacturer build lighter and smaller machines. Some of the advantages of an inverter are that it has smaller magnetic components (chokes and transformers) and it has higher efficiency and a faster response to the welding arc.

Inverter-based welding power sources operate at frequencies above 20 kHz, whereas traditional transformer power sources operate at a line frequency of 50 or 60 Hz.

Square-Wave Technology: Newer Welding Machines

This is the feature that applies to AC welding. A normal, older welding machine welds on AC current with sine wave for cleaning the aluminum weld surface. This means that the aluminum metal surface must have high frequency current imposed over the sine wave to clean the aluminum. A square wave welder can be adjusted for a longer time or a shorter, but more heat time to weld aluminum. Square Wave aluminum welding is much preferred over the older sine-wave welding.

Flow Meter

A flow meter allows you to monitor and regulate gas flow to the torch. Gas flow should be 10 to 25 cubic feet per hour (cfh). Any less would not *purge* the weld—displace all atmospheric contaminants around the puddle—allowing the weld to become contaminated. Much more than 30 cfh would be wasteful and could cool the weld too fast. Flow meters have built-in pressure regulators set by the manufacturer. There are flow gauges available that

For TIG-welding without high-frequency starting, use small copper strike plate for starting arc. Unlike steel, copper will not cause tungsten to break off when starting.

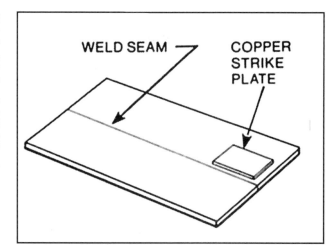

This Penske Racing welder is making a TIG weld on an exhaust collector for one of the Penske's Nextel Cup race cars. He is using one of the Invertec 205 T AC/DC combination machines. Note his plastic pipe welding rod holders beside the welder and in front of the argon bottle. Courtesy Lincoln Electric Co.

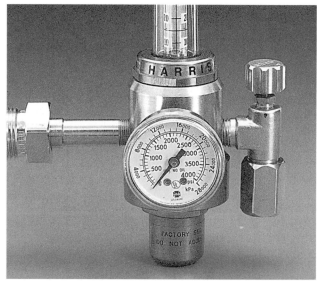

Keep a spare flow meter and a spare argon bottle in your shop for back-gassing stainless steel welds and titanium welds. Courtesy Lincoln Electric Co.

allow you to adjust both gas flow and pressure.

Cylinder—A cylinder is for storing a long-time supply of inert gas, usually argon.

Power Supply

The TIG power supply is similar to an arc-welder power supply. A TIG power supply may also include the following:

•Foot-pedal amperage control adjusts arc temperature and intensity. Rocking the pedal with your foot also starts and stops arc voltage. Pushing with your toe starts the arc and increases amperage for increased heat and weld penetration; pushing with your heel reduces amperage and stops the arc. It can also be used for stick-electrode welding. When using a foot control, I set amperage about 20% higher than what I think is needed. This way, I get extra heat by pushing on the foot pedal. I

don't have to stop the readjust the power-supply amperage control. Note: Stick welding with a foot pedal enables you to "back off" the amps at thin spots. Likewise, you can add amps where extra heat is needed. And stopping the weld with the foot pedal allows you to reduce heat at the edge of the workpiece and avoid a flat, thin bead common with conventional stick welding.

• Solenoids built into welder cabinet start and stop gas flow and cooling water. These solenoids self-time water flow to cool the torch and give timed post-purge of gas so the hot weld bead is not contaminated while cooling. Consequently, water and gas flow continue for a short time, even though the arc has been stopped. To take advantage of the post-purge gas, the torch must be held for a moment over the end of the weld bead after the arc has been stopped.

• A high-frequency power supply has a cleaning feature required for TIG-welding aluminum and magnesium. Its continuous spark actually cleans the puddle area by providing an electric-field shield over the weld bead. This is in addition to the feature that allows the arc to be started without the need to touch the tungsten electrode to the base metal. High frequency is shut off once the arc is established when DC current is used; it continues with AC current. A separate switch provides for start high-frequency or continuous high-frequency.

•Range and polarity switches select AC, DC straight or DC reverse polarity for any kind of welding.

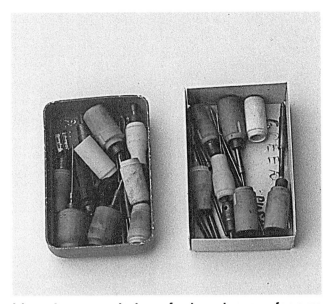

I keep two separate trays for tungsten: one for pure tungsten and one for 2% thoriated tungsten. The pure tungsten, cups and collets are kept in a green painted tray, the 2% thoriated tungsten, cups and collets are kept in a red painted tray so I won't mix them up, even when the color codes are ground off the tungsten. You may also want to have additional color coded trays for the other grades of tungsten, such as ceriated, lanthanated, zirconated, and tri-mix tungstens. See the sidebar on page 114.

• Portable base or cart contains necessary items such as gas bottle, water reservoir and foot-control pedal.

• Water reservoir is extra nice because it allows continuous or high-amperage welding without danger of overheating the torch. A good water reservoir holds about 5 gallons and includes a motor-drive pump. The water should be mixed with 50% ethylene-glycol antifreeze if below-freezing temperatures are expected. If you use antifreeze, change it periodically. Antifreeze degrades and gives off a noxious odor when stale.

TIG Spare Parts & Special Equipment

To avoid making too many trips to the welding supply store, keep a supply of TIG welder parts in your toolbox. A shopping list is located in the sidebar on page 113. In addition to those items, I recommend that you use a spare argon flow meter when you are welding stainless steel or titanium and need to back-gas the weld to prevent "sugar" on the backside of the weld. I also recommend a spare TIG torch with power and water hoses.

To supplement the deluxe TIG setup, you'll need a good arc-welding helmet, gloves and a small toolbox with trays for the spare parts. I have two small trays for tungsten—one for dull electrodes

TIG SPARE PARTS LIST

Number	Description
3	#10 ceramic cups
3	#8 ceramic cups
6	#6 ceramic cups
3	#4 ceramic cups
4	1/8" tungsten, 2% thoriated (for steel)
4	1/8" tungsten, pure (for aluminum)
4	3/32" tungsten, 2% *thoriated (for steel)
4	3/32" tungsten, pure (for aluminum)
4	1/16" tungsten, 2% thoriated (for steel)
4	1/16" tungsten, pure (for aluminum)
2	1/8" collets
2	1/8" chucks
2	3/32" collets
2	3/32" chucks
4	1/16" collets
4	1/16" chucks
1	stubby back cap
1	2" back cap
2	small stainless-steel wire brushes

*Thoriated means tungsten includes thorium alloy for making it easier to start the arc. Unfortunately, thorium also can contaminate aluminum and magnesium, so pure tungsten should be used for these applications. Read the sidebar for descriptions of other grades of tungsten.

Reduce trips to the welding supply store by keeping a supply of TIG parts in your toolbox.

and one for sharp electrodes. When the tray of sharp tungstens is depleted, I sharpen the dull ones and transfer them to the sharp tray. This saves time when I need to change electrodes.

In the toolbox, keep several pins, clamps and setup fixtures handy. An important tool is a 6" x 12" framing square. If you recall from the gas-welding chapter, the framing square is useful for setting up parts and checking for possible warpage. Most things we weld are square to something, so a square comes in handy. You'll also need:

• Files of various sizes and descriptions. As discussed in the gas-welding chapter, you should

TUNGSTEN DESCRIPTIONS

Pure Tungsten—Non-Radioactive
Color Code: Green
Non-radioactive with a low current capacity. Recommended for use in AC welding of aluminum only.

2% Thoriated—Radioactive Element
Color Code: Red
2% thoriated material is the primary tungsten used in the United States. It is primarily used in DC welding. It has a high load and amperage capability. Welding vapors, grinding dust and disposal of tungsten scraps raise safety and health concerns because thorium is radioactive.

TRI-MIX—Non-Radioactive
Color Code: Cream
Tri-Mix tungsten has most of the similar performance characteristics of 2% thoriated tungsten. Three oxides are used to promote mitigation and consistency and in some cases, improved welds. Tri-Mix usually exhibits an increase in service life as well as an increase in arc starts between sharpening and fewer misfires on starting.

Zirconated—Non-Radioactive
Color Code: Brown/White
Recommended for use in radiographic-quality welding, where tungsten contamination must be minimized. Not recommended for DC welding. Tends to ball up in AC welding.

2% Ceriated—Non-Radioactive
Color Code: Orange
Recommended for use in short welding cycles such as in DC orbital pipe, tube, thin sheet, and small part welds. Not recommended for higher amp welds.

2% Lanthanated—1.5% Lanthanated Non-Radioactive
Color Code: Blue/Gold
The most commonly used non-radioactive tungsten as an alternative to 2% thoriated.
It offers a long electrode life under heavy loads. Results usually include improved and more stable arc and long service life.

Specifications courtesy of Diamond Ground Products;
www.diamondground.com

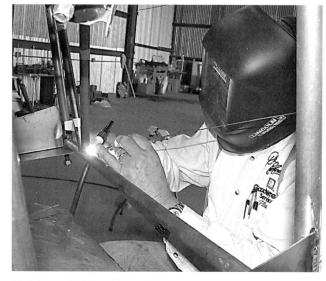

Making a TIG-weld repair to a thin wall chromemolly aircraft fuselage, using a foot pedal for amp control. The welder is not wearing gloves in order to get a beter feel of the TIG torch and the bare 4130 welding rod. The welding machine is a TIG 205 amp AC/DC unit. Photo by Jim Holder

have a coarse round file, a coarse half-round file and a flat mill file.

• Bench grinder with fine- and medium-abrasive wheels. Use these for sharpening tungsten and fitting small parts.

• Belt sander and disc sander for fine-dressing tungsten and fitting parts to be welded.

• Other niceties include a comfortable chair, clean welding table and good lighting. Mount a fluorescent light over your welding table. Two or three portable flood lights for welding such things as engine mounts, race car frames, trailers or airplane fuselages may come in handy. A clean, well-lit welding area is safer and will result in better welds.

Maximum TIG Setups

If money is not a problem, or if you want to set up a first-class race car or airplane shop, you should consider the top-of-the-line TIG machines. These come with computer, digital memory programs that you can repeat each time you do a specific welding job. And these machines can and do provide much better arc quality than the bottom-of-the-line welding machines.

Pulsed Arc—Years ago, welders had to pump the foot pedal to keep from melting holes in their parts. Now, the welding machine can be adjusted to give pulses of full heat with low amp power alternated in

The welder is sitting down to make it easier to use the foot pedal to adjust the torch heat and amps. His work would be more accurate if he had something to rest his wrists or elbows on to steady his hands. Photo. Jim Holder.

This is a water-cooled torch that has been adapted to work with an air-cooled TIG-welding machine. The brass DIN fitting at the right has a special water fitting that can be hooked up to work with a separate water pump that is portable. You will have to special order the fitting from a welding supply dealer.

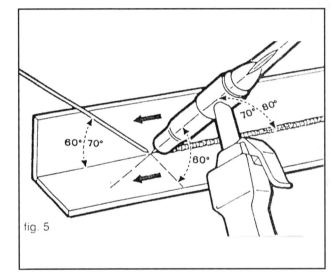

fig. 5

The correct angle for TIG welding in a corner is shown here. But always manipulate the torch angle to point the heat where required for good welds. Courtesy Daytona MIG Co.

between the full power pulses. Pulsed current welds allow the puddle to "freeze" between filler rod application so that better puddle control is possible. Ask your welding supply for a demonstration before you buy.

Minimum TIG Setups

DC Conversion—If you already have a DC arc welder, you can convert it to a TIG welder by adding an air-cooled torch, and an argon flow meter/flow gauge and tank. This will restrict your TIG welding to steel and stainless steel because of the lack of a high-frequency unit. Because of this, you'll also have to start an arc on the base metal or a copper strike plate. A strike plate makes arc-starting easier and reduces tungsten contamination from the base metal. Also, copper won't break the tungsten as easily as steel or aluminum, but it's not as good as high-frequency arc-starting. Without a foot-operated amperage control, you must stop welding to make amperage adjustments at the machine.

AC Conversion—If you have an AC/DC arc welder, you can fit a high-frequency unit to it—for about the cost of your AC/DC machine—a torch, hoses and regulator, and argon gas cylinder. This will allow you to TIG weld aluminum. Again, you won't have a foot-operated amp control. You'll also have to touch the tungsten to the base metal or strike plate to start the arc.

Whichever TIG setup you choose, you'll be able to weld things you never thought possible, such as welding thin parts to thick parts or complicated

assemblies without burning them up. But mainly, TIG is so squeaky clean that you can weld in your Sunday clothes and not get dirty. Although I don't recommend this, I've actually welded airplane parts in a suit, white shirt and tie!

Although slow compared to arc and MIG welding, TIG-welding is considered the most precise method of shop welding. Consequently, it's used to repair bad welds where accuracy counts.

Inverter Technology

In the 1970s and early 1980s, a good TIG machine weighed as much as 400 to 500 lbs. Today, because of inverter technology, a machine even more powerful and more precise can weigh less than 50 lbs. But this current (no pun intended) technology has allowed more features to be included in top-of-the-line welding machines with no penalty in size and weight.

Be aware of the newest welding machines that provide capacitor start, pulsed start, balanced and unbalanced wave forms, as well as asymmetric wave forms. All these features contribute to more precise machine settings and more precise welds. These machines will not be covered in this book because they deserve a complete book dedicated solely to their setup and operation. There is just too much to say about them to try to put it in this single chapter.

Heat-affected area of TIG weld seam is less than that for most other types of welding. Because of high temperature differential and small heat-affected area, avoid TIG welding in a draft

USING A TIG WELDER

To refresh your memory, an electric arc is generated between a non-consumable tungsten electrode and the base metal in TIG welding. Non-consumable means that the electrode is not intentionally melted into the weld puddle as in conventional arc welding. The tungsten will, however, erode and become contaminated in use. Consequently, it will also be ground away as you dress it to the desired point again and again.

The type of metal being welded determines the tungsten tip shape required. For instance, the sharp point needed to weld steel confines the arc to a smaller area, resulting in more concentrated heat at the weld seam. But a crayon-shaped point is used to weld aluminum because that metal dissipates heat more quickly and needs more area heated at the weld seam. This is done with the resulting broad arc. More on dressing tungsten tips later.

As the arc heats a molten puddle on the base metal, dip the filler rod into it as you would if gas welding. The inert gas from the torch shields the puddle from atmospheric contamination.

To ensure that shielding gas covers the weld while solidifying, keep the torch over the bead after completion until the purge gas stops. The timer on the TIG machine usually provides 5-6 seconds of gas flow after the torch is off. This keeps 4130 steel from cracking and helps prevent crystallization in stainless steel. Also, titanium demands post-purge gas to prevent atmospheric contamination of the weld bead.

Polarity Settings

Most DC TIG welding is done on straight polarity—electrode is negative. There is no polarity with AC current. AC current constantly changes direction. There are 60 complete cycles a second in the U.S. AC current frequency is 50 cycles per

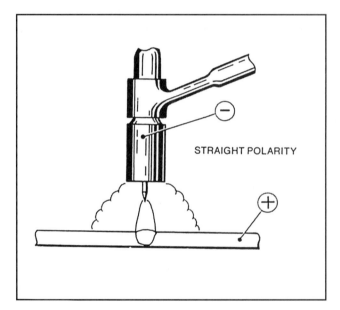

Set TIG welder to DC current, straight polarity for welding mild steel, 4130 steel, stainless steel and titanium.

second in Europe and 90 cycles per second in other parts of the world.

When welding most mild steel, stainless steel, 4130 steel and titanium, set your machine for DC current, straight polarity. This concentrates most of the heat at the work—about 70%. However, when welding aluminum or magnesium, use AC current. The reason for this is the oxide-cleaning feature of reversing current—very important when welding these non-ferrous metals.

As for TIG welding with DC current, reverse polarity—electrode is positive—you'll rarely have the occasion. The tungsten may melt before the base metal because about 70% of the heat is at the

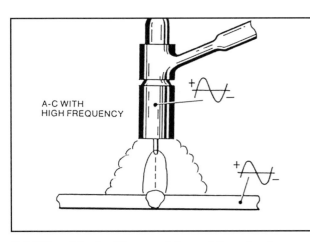

Set TIG welder to AC current for welding aluminum and magnesium.

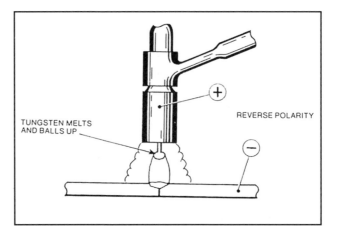

DC current, reverse polarity is a seldom-used TIG setting because tungsten melts before base metal. Higher heat is at electrode, not work. Therefore, its only application is for welding thin-gauge metal.

electrode, not at the work. Consequently, penetration of the base metal is poor. However, DC current, positive polarity does have a useful application. It is best used for welding very thin sheet metal—not aluminum or magnesium. Shallow penetration then becomes an advantage.

Tungsten Sharpening

I coarse-sharpen tungsten electrodes on a small bench grinder, then finish dressing them on a power sander with 120-grit paper. The grinder removes metal fast, and the sander does a good dressing job. Remember to grind tungsten slightly blunt for welding aluminum and sharp for welding steel, stainless steel and titanium.

A sharp tungsten tip is best for welding steel because it provides for better control and concentrated heat. But, because the heat for welding aluminum needs to be more evenly distributed and AC current melts the sharp electrode point—tungsten/base-metal heat distribution is about 50/50 AC—a sharp tip would contaminate the weld. The blunt tip cures both of these problems. It scatters the arc, distributing the heat, and it doesn't melt.

Don't go to the trouble of balling, or rounding the tip of the tungsten for aluminum welding as some welders do. It will ball itself in about two seconds of welding.

Sharpen tungsten lengthwise using a series of straight cuts toward the tip. Never sharpen tungsten by rotating it against a grinder. Sharpening a tungsten in this manner will result in a poorly controlled arc pattern. See drawings nearby.

Install & Adjust Tungsten—Loosen the collet on the back of the torch to free the tungsten. This will allow it to slide in and out. Adjust the tungsten tip so it projects about 1/8" to 1/4" past the tip of the cup.

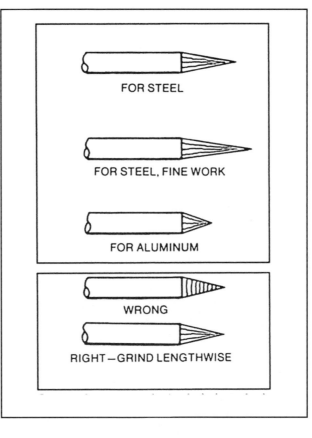

The shape of the tungsten electrode is important. Sharpen tungsten to look like a pencil for welding steel. Sharpen to needle point for fine work. Sharpen tungsten to crayon-like point for welding aluminum. Always grind tungsten lengthwise.

TIG Welding Tips

To make things easier, read the following before you start TIG welding:

• Once again, master gas welding before you attempt TIG welding.

• Clean the base metal as if you were going to eat off it. Seriously, you can't get the metal too clean for TIG welding. There is no flux to float off impurities.

Practice, practice, practice. These practice welds in aluminum and stainless were made with a Lincoln square wave, pulsed TIG machine.

This welder is making a very delicate weld repair to a part of the engine on the Penske Racing NEXTEL Cup race car. He is not using a foot pedal because he can't reach a pedal, so he is just setting the machine to a given amp and welding. Courtesy Lincoln Electric Co.

• Cut the welding rod into 18"-long pieces. This usually means cutting the 36"-long rods in half. Shorter pieces are easier to use.

• Tungsten diameter should be about half base-metal thickness. For example, use 1/6" diameter tungsten to weld 1/8" thick metal.

• Cup size should be as large as possible without restricting access to the weld. For instance, you'll have to use a smaller cup to weld in tight corners. Use a #8, #10 or #12 cup for flat steel and where access is good; #4 or #6 cup for corners and where access is poor.

• Clean the welding rod before you start welding. Use MEK or alcohol on a clean, white cloth. Even dust contaminates the weld.

• Make sure the lighting is good because the light from the arc is less intense than in other welding. Use a clean, #10 helmet lens for most TIG welding.

• Do not allow even the slightest breeze or draft in the weld area. Cool air will crack a TIG weld because the heat-affected area is smaller and more sensitive to rapid cooling. A breeze can also blow away the shielding gas.
•Never touch the hot tungsten to the puddle or filler rod. Capillary action causes molten metal to wick up—flow up—the tungsten, contaminating it. The weld is also contaminated as the wicked metal oxidizes and boils off the tungsten, blowing oxidized metal into the puddle. If this happens, stop welding immediately and grind the wicked metal off the tungsten. This is why you'll use more tungsten now than after you have more experience.

• Use a gas lens cup if you have room. The gas flow will be more uniform and your welds will look better.

• When—not if—you burn a hole through the base metal, completely let up on the pedal or break the arc. Let the puddle cool before continuing.

• Always shield the TIG weld light so you won't burn someone's eyes. TIG appears to give off less light, but its ultraviolet radiation is just as dangerous.

• Sit down while welding, if possible. Be comfortable. Take advantage of the fact that TIG welding generates no sparks to fly into your lap.

• Always tack-weld parts before running your final bead.

• Before making a critical TIG weld, try your procedure on a test specimen first.

• As in oxyacetylene welding, make a molten puddle, then dip filler rod into it. Don't try to melt

Jim Holder of Alamogordo, NM, is using an air rotary grinder and abrasive cut-off wheel to cut out some bent thin wall 4130 steel tubes that bent because of a poor design of his Nesmith Cougar Homebuilt airplane. When the bad tubes are removed, Jim will replace them with a better design and a new landing gear.

Here is the new landing gear that is replacing the damaged one on Jim's airplane. The new gear will put the loads in a vertical plane rather than in torsion as the original gear was. The new tubes have only been tack welded for fit-checking this far.

The tubes have been fitted and splice tubes are being tacked into place according to the FAA Publication AC-43- 13A, which I wrote in 1998. All aircraft welding repairs on planes under 10,500 lbs. MUST be repaired to this specification.

Here I am making the final welds to the new landing gear installation, but sitting down to get a better view makes it impossible to use a foot pedal, so I'm using a thumb-operated amp control to control the weld puddle. Photo by Jim Holder

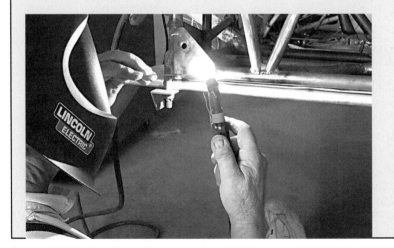

This close-up shows how the thumb-operated amp control is used to control the heat and therefore the weld puddle. Note the Lincoln Invertec 205-T AC/DC welding machine sitting on the toolbox in the background. It is plugged in to 220 volts in this photo but it will also work on 110 volts with no wiring changes. It weighs just 33 pounds. Photo by Jim Holder

My friend Larry LaBeau welded up a 4130 tube framework and covered it with sheets of printed circuit board to build this experimental Dyke's Delta airplane. Then he installed a Mazda Rotary engine and is now flying it! Quite a satisfying thing to do with his TIG-welding skills.

I welded this boat radar antenna mount from 316 stainless steel. 316 grade stainless is least susceptible to salt water corrosion. The radar antenna manually tilts to match the tilt of the yacht when under full sail.

Note the small 33-lb. Lincoln Invertec 205-T AC/DC welding machine in the background that is being used to weld this Nextel Cup race car at the Penske Racing shop. The crew member is TIG-welding a bracket on the engine of the car. Courtesy Lincoln Electric Co.

work. Later on, you should be able to judge a 1/4" to 1/8" arc gap without touching the cup to the work.

Get comfortable. Rest your arms—covered with long shirtsleeves—on the welding table or workpiece. Assume the most comfortable position that allows you to operate the foot pedal. Never weld with your arms unsupported. If possible, lean your shoulder against something steady.

Button up your collar to prevent neck burns from the ultraviolet light.

Keep the welding rod away from the arc until you have a molten puddle. After it has started, dip the rod in and out of the puddle as you move along the weld seam. Do not try to melt the rod with the arc. Let the molten puddle do the melting. Dipping the rod into the puddle cools the puddle slightly. So the rhythmic in-and-out motion of the rod maintains a constant puddle temperature. If dipping cools the puddle too much, compensate by increasing heat slightly.

Tip: This one is worth repeating, particularly for doing TIG welding. Drill a #40 hole in tubing when welding it closed. This will keep the puddle from blowing out onto the tungsten as you close the seam. Hot air expanding inside the tubing causes this.

The following are specifics on welding popular metals. Suggestions are given on how to set up the machine and provide for additional shielding, if needed. Except for magnesium, refer to Chapter 4, for information on specific alloys and the respective filler rods to use. Magnesium rods are covered in this chapter, page 126.

the rod with the arc. A cold weld, with poor penetration, results if you try to drop molten rod into the seam.

Ready to Weld

The torch should be in your right hand if you're right-handed. Switch hands if you're left-handed. To begin, rest the corner of the cup on the work at about 45°. Don't allow the tungsten to touch the

TIG-WELDING STEEL

Because mild steel and chrome moly are TIG-welded with the same process—machine and filler material—you can weld mild steel to 4130 or vice versa. For welding either, set up the machine by using the following procedure:

• Use steel welding rod #7018 or equivalent.

• Use DC straight polarity—negative electrode.

• Set high-frequency switch to START, if so equipped.

•Adjust gas flow to 20 cfh.

•Set water-cooling switch to ON, if so equipped.

•Set coarse- and fine-amp controls—coarse to 75 amps, fine to 80% for 0.063" steel. This gives 75 x 0.80 = 60 amps with the pedal all the way down. For thick metal, try these settings: coarse to 125 amps and fine to 35%, or 44 amps with full pedal. If this doesn't provide enough heat, increase the fine-amp setting. Go 50%, 80% or whatever heat you need. The reason for the lower fine-amp setting for thicker materials is to increase duty cycle. Regardless of the fine-amp setting, final amperage at the electrode is controlled by you at the pedal, or the thumb control.
•Turn on contactor and amp-control switches to remote foot pedal or thumb control, if so equipped. Contactor switch turns "on" the foot pedal.

TIG-WELDING STAINLESS STEEL

TIG-welding stainless steel is similar to welding mild steel and 4130 steel. However, you'll need some extra items: an extra bottle of inert gas, extra flow meter and about 20 feet of argon hose. This assembly of items will be used to purge, or shield, the back side of the weld.

Back-Gas Purging

As mentioned earlier, purging is the process of displacing all atmospheric gases and replacing them with an inert shielding gas such as argon. I usually set purge-gas flow the same as or about 25% more than torch flow.

The reason for using back-gas purge is that molten stainless crystallizes if it's exposed to air. Sugar, or crystallization, on the backside of the weld would weaken the weld and base metal considerably.

It's easy to back-gas stainless tubing such as an

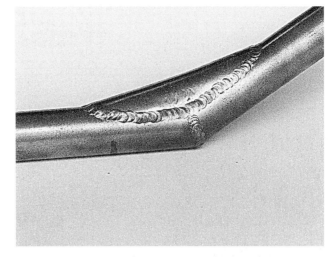

This otherwise excellent TIG weld on this aircraft 4130 steel tube mount is defective because the weld bead is concave and subsequently weak in one spot. Adding another weld bead on top of the low one would be acceptable.

engine exhaust pipe. You simply fill the tube with gas. Cap both ends of the tube with making tape, punch holes in the tape, stick the purge hose in one end and a short piece of open tubing in the other end to exhaust the argon. At 15-20 cfh gas flow, the average exhaust pipe can be purged in air in 4–5 minutes. Larger pieces, such as long, 12" ID tubing, take more time to purge.

Flat stainless plate is harder to purge. But you can build a cardboard-and-masking-tape cover over most seams to act as a purge shield. Simply enclose the backside of the weld seam with craft-type cardboard and masking tape. Shape the cardboard so it straddles, but doesn't touch, the weld seam. Burn a large hole in the cardboard and the purge gas will be lost. Cap the ends with masking tape and insert the purge and exhaust hoses. The accompanying drawing illustrates how to make a back-gas shield for flat plate.

TIG-WELDING TITANIUM

Although most alloys of this expensive, lightweight metal can be TIG welded with the same basic machine setup as mild steel and 4130 steel, it requires a more elaborate shielding-gas apparatus than stainless steel. As a result, some titanium alloys are not weldable at all. This is because of titanium's extreme sensitivity to contamination. Hot titanium reacts with the atmosphere and dissimilar metals, causing weld embrittlement. This contamination is serious if carbon, oxygen or nitrogen is present in sufficient quantities. In the solid state, as in a weld heated above 1,200°F (649°C), titanium absorbs oxygen and nitrogen from the air.

Simple back-gas shield for TIG-welding titanium can be made from cardboard. Masking tape is used to hold it to the workpiece. Use the same kind of shield for welding stainless steel and 4130 steel.

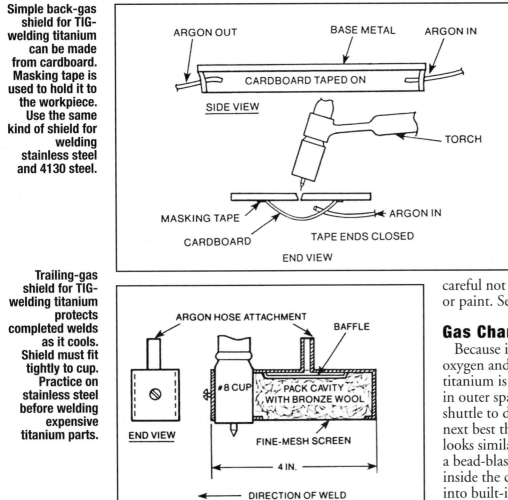

Trailing-gas shield for TIG-welding titanium protects completed welds as it cools. Shield must fit tightly to cup. Practice on stainless steel before welding expensive titanium parts.

An argon-gas shield must cover both sides of a weld seam while titanium is at the 3,263°F (1,795°C) molten stage and all the way down to 800°F (427°C). Otherwise, embrittlement from contamination and resulting cracking will occur.

Make a device to provide a trailing shield to cover the weld until it cools to below 800°F (427°C). Usually, after the trailing shield passes over the completed weld, the titanium has cooled sufficiently. You can also use temperature-indicating crayon or paint to be sure. But if you do, be careful not to contaminate the weld with the crayon or paint. See the drawings of trailing-shield device.

Gas Chamber

Because it is extremely reactive with nitrogen, oxygen and hydrogen, the best place to weld titanium is in a total inert atmosphere such as that in outer space. So, if you can get a ride on the space shuttle to do your welding, great. Otherwise, the next best thing is a gas chamber. Such a chamber looks similar to an incubator for newborn babies or a bead-blasting cabinet. Put the part to be welded inside the chamber, close it, then put your hands into built-in gloves to do the welding. The chamber is completely purged with argon so no air exists. Much purer titanium welds will result.

KNOW BEFORE YOU START

Because welding titanium is so specialized and the technique so critical, you should discuss your titanium-welding project with a welder who specializes in titanium before attempting it. If you can't find such a person, RMI Company, Niles, OH 44446 is a company that specializes in welding titanium. They would be glad to answer your questions. Or contact the American Society for Metals, Metals Park, OH 44073. Their educational division, the Metals Engineering Institute, can provide instructional material through their Home Study and Extension Courses for all types of welding techniques and metals. Finally, there's the American Welding Society (AWS), 2501 Northwest 7th Street, Miami, FL 33125. Their four-volume Welding Handbook will give you all you ever wanted to know about welding and more.

GAS SHIELDING FOR TITANIUM WELDING

Material Thickness	Torch Argon Flow (cfh)		Trailing Shield Argon Flow (cfh)	Back Gas Helium or Argon (cfh)
0.030	15	15	3	
0.060	15	20	4	
0.090	20	20	4	
0.125	20	30	5	

TIG-WELDING ALUMINUM

TIG-welding aluminum is slightly different than welding steel. Machine settings are different. And, as you may remember from the gas-welding chapter, aluminum doesn't change color as it forms a puddle—it gets shiny instead. However, unlike gas welding, flux isn't needed to TIG-weld aluminum. In fact, flux would really mess up the weld and your TIG welder. The following tips will help you TIG weld aluminum:

• The weld seam area should be as clean as possible, and it should be free of aluminum oxide. Remove the oxide immediately before you weld—not a week or two before—by mechanical or chemical cleaning. Mechanical cleaning is done with a stainless steel wire brush, sandpaper or abrasive pads. Afterward, wash off the dust with soapy water and rinse with clear water. Then, wipe down the seam area with alcohol, MEK or acetone.

• To eliminate the need for cleaning newly fabricated aluminum parts, use the more expensive paper-covered sheet stock. Adhesive-backed, paper-covered sheet can be cut, formed and fitted with the covering in place. When you're ready to weld, simply peel back the covering from the weld seam to expose super-clean aluminum, and you're ready to weld.

• If TIG-welding aluminum castings or forgings, V-groove the joint all the way through. Or V-groove the joint from both sides and weld on both sides, if possible.

• If welding aluminum plate that's more than 1/16" thick, V-groove the joint for better penetration.

• Don't attempt to weld 2024 aluminum or other non-weldable aluminum alloys. Read on for a handy test to check the weldability of unknown aluminum alloys. Weldable alloys include 1100, 5052, 6061 and all castings.

Machine Settings
• Polarity selector switched to AC.
• High-frequency unit set to CONTINUOUS—an

GUIDELINES FOR TIG-WELDING ALUMINUM

Material Thickness (in.)	Current (amps)	Tungsten Diameter (in.)	Weld Rod Diameter (in.)	Argon Flow (cfh)
0.020	25	0.040	1/32	16
0.040	34	1/16	1/16	18
0.063	50	1/16	1/16	20
0.080	75	3/32	3/32	20
0.100	100	3/32	3/32	22
0.125	125	3/32	1/8	25
0.250	150	1/8	1/8	35

Chart gives approximate settings for welding aluminum. Adjust amps to suit the particular conditions.

ALUMINUM HEAT TREATMENT NOMENCLATURE

T0: Completely soft, no temper
T2: Annealed by heat to soften (cast only)
T3: Solution heat-treated, then cold-worked
T4: Solution (salt bath) heat-treated
T5: Artificially aged.
T6: Solution bath heat-treated, then artificially aged
T7: Solution heat-treated, then stabilized
T8: Solution treated, cold-worked, artificially aged
T9: Solution treated, aged, then cold-worked

---

absolute necessity.

• Argon flow meter set to 20 cfh, or check flow chart.

• Water cooling switched ON, if so equipped.

• Contactor and amp switches on REMOTE, if equipped with foot pedal.

• Set coarse amp adjustment to about 60 amps; fine amp adjustment to 70%. Reset amp setting(s) if heat is not right.

Also, you'll need to use the following items:

• Pure tungsten rather than 2% thoriated tungsten for TIG welding aluminum. As you may recall, thorium contaminates aluminum welds.

• Aluminum welding rod 4043 for most alloys. Check the chart to be sure

ALUMINUM ALLOYS			
Alloy No.	Tensile Strength (psi)	Heat Treatable	Weldable
1100	15,000	No	Yes
3003	26,000	No	Yes
3105	23,000	Yes	Yes
5005	26,000	No	Yes
5052	41,000	No	Yes
5086	47,000	No	Yes
2024	61,000	Yes	No
6061	42,000	Yes	Yes
7075	65,000–75,000	Yes	No

Note: If welding becomes necessary after part is heat-treated, heat-affected area is likely annealed or softened. If heat-affected area is small, it usually does not require heat-treating again. But if affected area is large, part should be heat-treated again. Refer to chart to determine if aluminum alloy is weldable.

Griffin Aluminum Radiators exhibit excellent TIG-welding. Try to make your aluminum TIG welds look this good.

Preheating

Not too many years ago, I toured one of the largest aircraft engine repair shops in the world. I was shown the welding shop where cracked and broken cast-aluminum cylinder heads were repaired. They used a natural gas oven to preheat three or four cylinder heads at a time before welding. The cylinder head was heated evenly to 350°F (177°C).

Any aluminum more than 1/4" thick benefits from preheating before welding. A simple way to preheat is in a gas or electric oven with thermostatic control, such as a kitchen oven. Just make sure the aluminum you're welding is absolutely clean, or you'll stink up the house for a week!

It's usually not necessary to re-preheat during welding, because the weld heat keeps the part hot enough.

Weldability

Several aluminum alloys are not weldable, most notably 2024. Usually found in sheet form, this alloy is commonly used on airplane-wing and fuselage skins, wing ribs and fuselage bulkheads. If you're not sure about the alloy, look for previous welds. If it was welded before, it can be welded again. But if there is no sign of welding, try a sample weld to test it. Finding a part to do sample welds on might not be easy, but I would even go to the trouble of finding an unusable part to practice on to avoid ruining a repairable part. Race car shops and airplane shops usually have scrap parts in a corner somewhere. So, if an aluminum part is not weldable, repair it with rivets or nuts and bolts, or replace it.

Grounding Aluminum

When TIG-welding aluminum on a grounded tabletop, it usually arcs between the table and the work. Consequently, you may make a beautiful weld on a part that took all day to cut, shape and fit, only to turn it over and find arc burns and craters where it contacted the table.

Arcing can be prevented by grounding the workpiece directly or providing a simple ground for the work. I usually lay a heavy mechanical finger on the part to help hold it against the weld table. If you use a separate ground cable, make sure it's welding-cable size. A small-diameter cable will overheat from high amperage.

Weld Craters

When running an aluminum weld bead, don't break the arc or rapidly shut off the arc with the foot pedal as you reach the end of the seam. This will cause a small depression or crater at the end of the bead. Instead, back off the foot pedal slowly, then lead—move—the arc back to the already solidified bead. This will "freeze" the puddle while it's still convex.

Fitting Parts

Fitting parts closely is crucial when welding aluminum. Refer to Chapter 6. You can fill a gap, but the backside of the weld will look like sand or even gravel because the molten aluminum solidifies too quickly and forms a lumpy bead.

Back-Gas Purge

TIG-welding aluminum does not require a back-

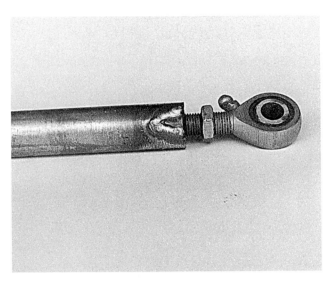

Race car suspension member is fishmouthed and TIG-welded.

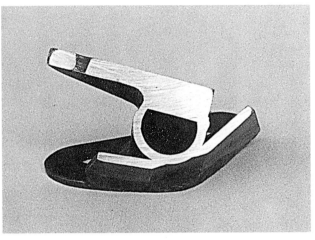

Don't try to TIG-weld the backside of your aircraft or race car parts unless the welding engineer specifies welding on both sides. This TIG-welded aircraft turbocharger bracket shows what cross section you should expect to see in your welds. The bracket has been sectioned to certify my production welds. The test passed aircraft certification.

gas to purge oxygen and hydrogen from the weld as with stainless steel and titanium. Although it can improve the appearance of the back side of the weld bead, it will not noticeably improve weld integrity. Aluminum does not pick up atmospheric contamination as does stainless steel or titanium. And, aluminum will solidify in open air with nothing more than oxidation of the surface. Consequently, using back gas to weld aluminum is a waste of valuable time, money and equipment.

Heat-Treating After Welding

Aluminum is very easy to shape, form and TIG weld in the dead-soft—O—condition. After these operations are completed, it is then heat-treated to give the soft aluminum strength and rigidity.

The 6061-0 alloy is commonly used to make fuel-tank bulkheads, wing ribs and other parts for airplanes and race cars. The complete assembly or subassembly is then heat-treated in an oven or brine—salt—solution. In the oven process, 6061-0 is heated to 950°F (510°C) for about 15–30 minutes and then air-cooled.

The brine solution process involves heating the solution to 1,000°F (538°C), a temperature at which it does not boil. The 6061-0 alloy is immersed in the brine for 15–30 minutes. It is then immediately quenched—cooled—in 70°F (21°C) water. At this point, the aluminum isn't completely hard, but after sitting 24 hours at room temperature, it age hardens to full strength and hardness. Age-hardening is also called precipitation hardening.

Note: After 6061-0 is heat-treated, it is transformed into either 6061-T4 (solution heat-treated)or 6061-T6 (oven heat-treated).

The chart on page 123 details heat-treatment nomenclature. For example, for aluminum-alloy 6061-TX, the T means temper, the following number indicates heat-treatment type.

Annealing—Annealing aluminum is the process of heating it to 750°F (399°C) and allowing it to air-cool slowly to remove the effects of previous heat-treatment so it can be cold-formed without cracking. It can be reheated after annealing.

TIG-WELDING MAGNESIUM

Magnesium can burn and support its own combustion. Water or dry-powder fire extinguishers will not put out a magnesium fire. In practical terms, the only way a magnesium fire can be extinguished is to wait for all the magnesium to be consumed. I once saw a race car with a magnesium transmission case and wheels burn to the ground. All that remained of the transmission were the axle shafts, gears, nuts and bolts. The magnesium parts burned to ashes. The intense heat melted the aluminum engine, leaving crankshaft and connecting rods laying in the dirt. A fully equipped fire truck was not able to extinguish the fire.

So, when welding magnesium, try to do so outside, away from flammables. Magnesium isn't likely to catch fire unless there are magnesium filings nearby—such as those created by machining or dressing the weld seam. If magnesium does catch fire, stand back and let it burn. You probably can't stop it.

Clean Before Welding

As with other metals, magnesium should be cleaned of all scale and corrosion in the weld-seam area before TIG welding. Use some aluminum wool, steel wool or a stainless-steel brush to remove the white, powder-like corrosion.

If the corrosion can't be removed by mechanical means, use chemicals. Mix 24-oz chromic acid, 5 1/3-oz ferric nitrate and 1/16-oz potassium fluoride in 70–90°F (21–32°C) water to make a 1-gal. chemical cleaning solution. Dip the part in this solution for three minutes, then remove it and rinse in hot water. Let it air-dry before welding. Don't use compressed air for drying. Compressed air may be contaminated with dirt, water and oil.

Stress-Relieving Magnesium

Magnesium alloyed with aluminum is susceptible to a unique phenomenon called stress-corrosion cracking. For example, if you drill a hole in magnesium and put in a tight-fitting bolt, the area corrodes, then cracks as a result of the corrosion. Check with the manufacturer to determine the alloy content of a particular magnesium alloy. Otherwise, you may need to have the magnesium analyzed metallurgically—an expensive process. These alloys must be heat-treated to remove the welding stresses, which would otherwise result in corrosion and cracking. See the chart below for stress-relieving by heat-treatment.

GUIDELINES FOR TIG-WELDING MAGNESIUM

Material Thickness (in.)	Current (amps)	Tungsten Diameter (in.)	Weld Rod Diameter (in.)	Argon Flow (cfh)
0.040	35	1/16	1/16	12
0.063	50	1/16	1/16	12
0.080	75	3/32	3/32	12
0.100	100	3/32	3/32	15
0.125	125	1/8	3/32	15
0.250	175	1/8	1/8	20

Listed are approximate values for welding magnesium. Make adjustments for conditions. Follow above recommendations for setting up to weld magnesium.

STRESS-RELIEVING MAGNESIUM THROUGH HEAT TREATMENT

MAGNESIUM SHEET			MAGNESIUM CASTINGS		
Alloy	Temp (F)	Time (min)	Alloy	Temp (F)	Time (min)
AZ31B-0	500	15	AM100A	500	60
AZ31B-H24	300	60	AZ63A	500	60
HK31A-H24	600	30	AZ81A	500	60
HM21A-T8	700	30	AZ91C	500	60
ZE10A-0	750	30	AZ92A	500	60
ZE10A-0	450	30			
ZE10A-H24	275	60			

To prevent cracking, magnesium sheet and castings must be stress-relieved by heating to given temperature and held there for times given. Heating is best done in an oven.

Chapter 14

PLASMA CUTTING

Penske Racing is using a plasma cutter to trim part of the frame on a new Nextel Cup race car. The plasma cutter will remove the metal with as little grinding as possible. Just picking up the torch and turning on the machine and air is all it takes to start a plasma cut. The crew member is only wearing tinted shop safety glasses. Courtesy Lincoln Electric Co.

Plasma cutters are extremely handy for making quick, clean cuts in any metal, including brass, copper, stainless steel or aluminum. Plasma is a gas heated to an extremely high temperature and ionized so it becomes electrically conductive.

BASIC OPERATION

In the basic plasma-arc process, invented and developed by Linde, an electrode is located within a torch nozzle. This nozzle has an arc-constricting orifice. Inert gas, usually argon nitrogen, or just dry shop air, is fed through the nozzle, where it is heated as high as 50,000°F (27,760°C), the plasma temperature range.

Because of the modest amount of skill required, plasma-arc is widely accepted for cutting ferrous and non-ferrous metals. The plasma arc's straight, narrow, column-like shape and high-current density mean that it is not critical to maintain a certain nozzle-to-workpiece distance to obtain weld or cut consistency. Also, greater nozzle-to-workpiece distance is possible with the plasma techniques mean better visibility of the workpiece for controlling the puddle or cut. However, the best cuts occur when the nozzle is 1/8" to 1/4" from the work, and less torch contamination occurs when the torch does not touch the work.

PLASMA-ARC CUTTING

Plasma-arc cutting uses a highly constricted, high-velocity arc that penetrates the metal. However, up to 50,000 volts is used to melt the metal. Either compressed air from a shop air compressor or a blended, inert shielding gas is used to blow the molten metal out of the kerf. Because it uses a narrow, straight, column-like arc, there is minimal kerf width. And, because of the clean cut, the cut surfaces do not generally require cleanup. Metal up to 6" thick can be cut with a plasma-arc setup, depending on the type of metal and the arc current used.

Plasma arc can be set up to cut with nitrogen as its shielding gas. Although shielding gas is preferred for clean, no-oxidation cuts, shop air is also used because of its low cost. Because high-pressure, high-velocity shop air is extremely noisy, it can be distracting. The sound is similar to listening to a compressed-air blow gun at close range.

The plasma-arc cutting torch is excellent for autobody shop work because it will cut through paint, undercoating, body putty and dirty metals. No precleaning is necessary. Because it doesn't use the oxidizing process to cut metal, plasma arc is ideal for cutting high-strength, low-alloy steel used in new unibody cars. This non-oxidizing feature also makes it a natural for cutting stainless steel; and non-ferrous metals such as aluminum, copper and brass.

The big advantage of plasma-arc cutting is speed—up to 50 times faster than oxyacetylene. Available now are portable units complete with gas, torch and power supply, ready to plug into 220-volt, 60-cycle, single-phase power, and there are units that plug into 110-volt, 60-cycle power. There are also plasma cutters that can be used on 220 volts and then be plugged in to 110 Volt power with no wiring changes. They figure out which voltage they are plugged into !

PLASMA-ARC CUTTING

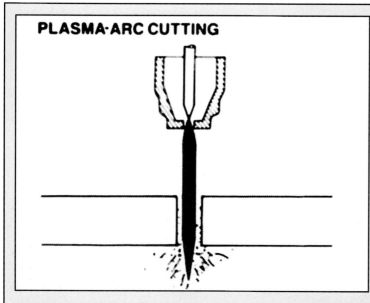

Plasma-arc cutting requires only high-energy gas column for cutting. Shielding gas is not required. Constricted-gas jet produces narrow, smooth kerf. Drawing courtesy Linde Welding Products.

These crew members are rebuilding a NASCAR race car and are using a Lincoln Electric Co. Pro-Cut 55 plasma cutter to make the repair job faster. This unit only weighs 55 lbs., making it easy to transport between track and shop. Courtesy Lincoln Electric Co.

The Penske Racing crew member is fabricating a new NASCAR race car and is using a Lincoln Electric Pro-Cut 55 plasma cutter, which only weighs 55 pounds. With a plasma cutter, there is no smoke or flame to contaminate the shop air. Courtesy Lincoln Electric Co.

John Gilsdorf of Alamogordo, NM, does a fit check of a replacement panel from an aftermarket supplier. The panel will need trimming before it can be welded into place.

John makes a plasma cut on the sheet metal replacement part to get it ready to MIG weld on the '56 Ford. Plasma is the perfect tool for auto body repairs.

John does a fit check of the new plasma trimmed sheet metal part that is now ready to weld into place. A lot of metal was trimmed off before it was fitting properly.

High School student Anthony Torres programs his laptop computer to make a plasma cut of an elk head while his dad looks on. Courtesy Torres Welding, Las Cruces, NM.

Here is the plasma cut on a sheet of 3/16" steel plate that shows the elk head that Anthony programmed on his laptop computer. Anthony apparently likes Chevys too. Courtesy of Torres Welding, Las Cruces, NM.

This plasma cut-out decoration was done by computer, but it is possible to plasma cut this design out by hand if you're experienced. Courtesy Pueblo Pipe and Steel, Alamogordo, NM.

This little Lincoln Electric Pro-Cut 25 plasma cutter only weighs 29 pounds and it will cut with either 110 volts or 220 volts and shop air. It was used to cut out some rusty panels on this 1964 Corvair Monza convertible.

The Lincoln 110–220-volt plasma cutter was used to cut out the rusty floor in the Monza convertible. The spray bottle contains phosphoric acid to slow down the rust in the floor metal after the trimming is done with the plasma cutter.

Jason is using the plasma cutter to trim out the replacement panel to fit into the floor of the '64 Monza convertible.

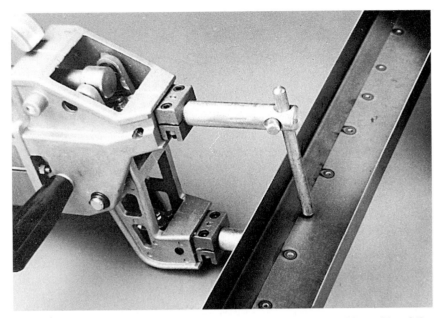

Steel flange spot-welding. Note the previous spot welds on either side of the spot welder tongs. Photo: H.T.P.

Chapter 15

SPECIAL WELDING PROCESSES

Other special welding processes used by hobbyists, craftsmen and production shops are: steel spot welding, aluminum spot welding, metal spraying, jewelry brazing and welding, metal shaping, abrasive cutting and—believe it or not—epoxy welding. Yes, there are times when welding heat is not appropriate, or when disassembly of the cracked block or oil pan would take too much time, and in certain selected cases, epoxy repairs are the answer. But first, we will discuss spot welding.

SPOT WELDING

Used strictly for joining one sheet-metal panel to another, spot welding is one of the oldest production-welding methods. The primary function of a spot welder is to make many welds with little effort and in a short time. It is also clean and requires no filler. Spot welding has been used extensively since the introduction of unibody—unitized body/frame—cars. It is also the standard welding procedure used by most sheet-metal fabricating industries.

This is one welding technique that can be understood and done without first learning how to gas weld. All that's necessary to produce a spot-weldable part is to have two clean pieces of sheet metal that will lay flat against each other at the weld joint. You cannot spot-weld a butt joint or T-joint because these joints lack sufficient surface area to clamp the two together. Therefore, all spot welds use lap joints.

In many cases, a lap joint is made by forming what is termed a *weld flange*. For example, if a panel butts into the side of another panel, as is the case of T-joint, a flange about 1/2" wide is formed at the butting edge of the panel to provide an overlap—lap joint—with the other panel. This overlapping flange, or weld flange, can then be spot welded.

You see such flanges every time you look under a car. It's the flange that runs lengthwise of the rocker panel, joining it to the floorpan and inner rocker panel.

Types of Spot Welders

Spot welding, more appropriately called *resistance welding*, uses pressure and electrical resistance through two metal pieces as the main ingredients of the weld. Heat is produced by high amperage routed through two mating workpieces clamped between two electrical contact points. No filler, flux or shielding is used. Because the pieces must be clamped together, access to both sides of the joint normally is required. Heating is local, usually a 1/4"-diameter spot.

Spot welding can be done to almost all steel and aluminum alloys, but equipment cost varies widely. Whereas a spot welder capable of welding steel costs less than a buzz-box arc welder, a spot welder for welding aluminum usually costs more than $50,000!

Although spot welding is simple by itself, complete spot-welding setups can get extremely complicated. The degree of complexity depends on how the welding unit is mounted. For instance, robot-operated spot welders are in wide use, particularly in the auto industry. However, for the small fabrication of a piece in a home shop, a pedestal-mounted spot welder requires that you take the work to the machine.

Regardless of the mounting, most spot welders operate on single-phase AC power. A step-down transformer converts the power-line voltage to about 250 volts.

Single-Phase Resistance—The standard, sheet metal shop spot welder is a single-phase resistance-type machine. With this type, AC current is passed to two copper electrodes that clamp two thin pieces of metal together. The resulting short

Four primary cycles of a spot weld are: squeeze, weld, hold and release. Pressure and current must be precisely controlled to assure good welds. Drawing courtesy The James F. Lincoln Arc Welding Foundation.

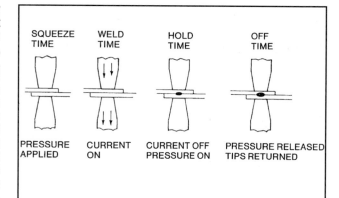

circuit causes the metal to heat to the melting point where the two electrodes meet. This melts and forces the panels together, locally, or at a spot.

Three Phase Rectifier—This welder consists of a three-phase step-down transformer with diodes connected to the secondary circuit. These water-cooled, silicone diodes are connected in parallel. Current is also passed through two clamping electrodes.

Capacitor Discharge—With the capacitor-discharge welder, which also uses clamping electrodes, a bank of capacitors is charged from a three-phase rectifier and then discharged into an inductive transformer. It can be used to spot weld dissimilar metals or delicate electronic parts.

Single Electrode—This is a simple unit that looks like a pistol and relies on a portable arc welder for its power source.

Unlike other spot-welder designs, the sheet-metal panels are not clamped together. This design features make it suitable for "skinning" sheet-metal

In spot welding, heat is produced by electrical resistance between copper electrodes. Pressure is simultaneously applied to electrode tips to force metal together to complete fusing process. Spot weld nugget size is directly related to tip size. Drawing courtesy The James F. Lincoln Arc Welding Foundation.

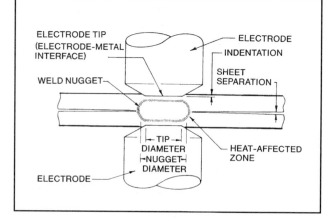

panels on automobile bodies where access to the inner panel is blocked. The primary appeal of this light-duty spot welder is for the auto hobbyist.

Aluminum Spot Welder—If you try welding aluminum with a spot welder meant for welding steel, it will only burn the metal. And it won't fuse

the pieces together. Aluminum spot welders are sophisticated devices programmed to slowly apply pressure to the spot weld while gradually increasing voltage and current. They develop up to 76,000 amps to spot-weld 1/8" thick aluminum.

Other Spot Welders—There are several other types of resistance welders, but most are high-production factory units, unsuitable for small workshops. For instance, projection welding utilizes small dimples, or projections, in the sheet metal. These projections arc as they touch another panel to complete the welding circuit, flash-butt welding the two together. A high electrical resistance at the projections is created, causing the panels to fuse together.

Seam Welding—Another type of resistance weld is called seam welding. Two pieces of sheet metal are drawn between two rollers while pressure and electrical current are applied to the knurled rollers. This action provides for a continuous resistance weld which is leak-free. A common application is in the manufacture of automotive gas tanks. The top and bottom halves are continuously welded at a flanged seam.

Most spot-welder designs are complex even though using them is relatively easy. Spot-welder settings for current applied, on-and-off cycle timing, electrode pressure, and the shape and condition of the electrodes all affect weld quality. Once the proper setup is achieved, the operator can make numerous welds of equal quality.

Aluminum Spot Welding

When spot welding low-carbon steel, a good weld can be made with a generalized setup using a wide range of current adjustments and an even wider range of clamp, current-applied, current-off, pressure-held, and pressure-released cycles or steps. However, aluminum is much harder to spot weld. It requires a precise clamping cycle and slow application of an initially low current, building up to a spike of high current, then tapering off to low current again. The clamping force is initially high to prevent arcing of this high-resistance metal, then backed off to prevent thinning of the aluminum, and finally increased as the weld cools. These steps must be carefully done in an accurate timing mechanism. And, as you may recall from the TIG and MIG chapters, aluminum is highly susceptible to surface oxidation that adversely affects weld quality. Consequently, aluminum must be absolutely clean to spot weld.

Also, the spot-welder contacts must be dressed—filed—often to keep them clean and to maintain electrode shape and size. Tip size is important. For example, when 31,960-psi pressure is applied to a

1/4" diameter surface, reduction of this contact-patch size to 3/16"-diameter will reduce the applied pressure to 17,978 psi—a 44% reduction! This will ruin the weld.

Because of the critical factors and complex machinery involved, aluminum spot welders can cost more than an entire welding shop equipped with MIG, TIG, arc and gas welders!

Not to worry, though. There are other ways to join sheet aluminum. For instance, at Aerostar Airplanes I used an aluminum spot-welding machine to weld two-piece landing gear doors. When the spot welder broke down, we substituted rivets for the spot welds—rivets and spot welds are similar in strength. Although spot welding is usually faster and much less prone to working, or moving, of the two sheets of metal joined, solid rivets will work nearly as well.

Weldability of Metals by Spot Welding

Almost any metal that can be fusion welded can be spot welded. However, oil, grease, dirt or paint on either or both of the parts to be spot-welded can have a negative effect on weldability. For this reason, it is imperative that you clean the sheet metal thoroughly prior to attempting spot welding.

Galvanized steel is a natural for spot welding. Galvanized coating was once used extensively in heating and air-conditioning ducting. It is now used in automobile bodies for rust prevention. Galvanized steel spot-welds easily with no special preparation to the metal. However, you should use a good breathing respirator because it generates poisonous zinc-based gases.

Stainless steel can be spot-welded, but it should be silver-brazed or TIG-welded instead.

Nickel-plated steel, such as that used in car-body trim, can be spot-welded. Unfortunately, heat will discolor the plating.

Refer to the weldability chart on the next page. It can be helpful for determining which metals can be spot-welded.

Metal Coatings—Special weld-through sealers are used in the auto industry for rust protection. They prevent water from entering blind areas in the body through weld seams. The sealer is applied to the facing side of a weld-seam half, then the panels are put in place and the seam is spot welded right through the sealer. This makes a watertight spot-weld seam.

How to Use a Basic Spot Welder

Most simple spot welders have four controls:
•On-off power switch.
•Foot pedal or hand lever to cause the two

electrodes to come together and make the spot weld.
•Electrode contact-pressure adjustment, usually air pressure.
•Amperage and timer adjustment. As metal thickness increases, so does arm-contact time and amperage.

Before making final spot welds, cut out some test scraps of the same material, about 2" square. Hold them together with locking pliers while spot-welding.

Start off with low pressure and amperage. Metals that have high electrical resistance, such as aluminum, require less overall current, but more exact control of the cycle. Steel and other metals with low resistance require greater amounts of current. Keep raising the pressure and amperage until the spot is about 3/16" to 1/4" diameter.

For instance, a spot welder with 1/4" electrodes should be set to the following values to assure good welds in steel sheet: 9800 amps, 32,000-psi ram pressure. Clamp time and weld dwell time are programmed into the welder. Make few spot welds using the suggestions just given and adjust the welder as required for good welds.

The distance between each weld is pitch. For most applications, weld pitch should be about 1" to 1 1/2". Judge the spacing as you move along a weld seam, making one weld after another. After making a few spot welds, perform the following tests.

Spot Weld Testing

The ultimate method of checking spot-weld quality is destructive testing whereby the two spot-welded pieces are pulled apart. However, you can tell a lot about a spot weld by its appearance.

Good spot welds are essentially round. The depression matches the size of the spot-welder electrode-contact points, with only a slight amount of heat-affected metal around the spot. The amount of indentation in the weld should be slight.

Bad spot welds look much different. Severe discoloration around a spot weld indicates overheating. Either the current was too high or the dwell time—duration—of the current was too long.

Little or no contact spot indicates a poorly fused spot weld. The current was too low or current dwell time too short. Spitting and sparking around the spot weld indicate one or more problems. Either the metal was dirty with paint, oil or grease, or the current was much too high for the metal thickness. It is also possible that clamping pressure was insufficient.

As with anything, it is easier to make corrections to spot-welding machine settings when you know

METALS THAT CAN BE SPOT-WELDED TOGETHER

	Aluminum	Stainless Steel	Brass	Copper	Galvanized Steel	Steel	Monel	Tin	Zinc
Aluminum	X					X	X		
Stainless		X	X	X	X	X	X	X	
Brass		X	X	X	X	X	X	X	X
Copper		X	X	X	X	X	X	X	X
Galvanized Steel		X	X	X	X	X	X	X	
Steel		X	X	X	X	X	X	X	
Monel		X	X	X	X	X	X	X	
Tin	X	X	X	X	X	X	X	X	
Zinc	X		X	X				X	

X indicates combinations that can be spot welded.

what caused the problem. Be sure to analyze the first two or three spot welds before continuing to make welds.

Destructive Testing—The most common way to check the operation of a spot welder is to perform a pull test. Spot-weld two metal test strips together. Start by visually inspecting the spot weld; then tear the two strips apart. Look at the torn-apart spot weld to determine whether it is good or bad. The amount of force it takes to tear the strips apart is the best indicator. If you have to really wreck the metal to tear the weld apart, the weld is probably good.

To tear the strips apart, clamp one strip in a vise. With pliers or vise grips, peel the other strip away from the first strip. After you have torn the strips apart, look at the weld area. Usually, there will be a hole in one piece of metal and a weld nugget—fused metal spot—stuck to the other piece. Pulling a nugget indicates the weld was stronger than the base metal.

Check the size of the nugget. It should be almost the size of the face diameter of the electrode. Check the shape of the nugget. It should have some of the metal from the other test strip attached to it. Look for brittle, crystalline-looking metal in the nugget. Brittleness is usually caused by excess heat, not enough pressure, or too much heat time. Look at the nearby drawings for how to do a spot-weld pull test.

Using a Single Spot Welder

If you are restoring an antique car, or building a new car and don't want to use rivets, a hand-held spot welder such as the Braze'n'Spot Welder from The Eastwood Company (www.eastwood.com), Malvern, PA, may give you the desired results. The advantage of this type of spot welder is that you can weld panels where access to the backside is restricted. Eastwood's spot welder operates off a 50-amp buzz-box/transformer and plugs into 110-volt house current.

To use it, first ground the welder to the work. Next, push the welding gun against the outer panel, forcing it against the inner. Do this while you retract the carbon electrode by pulling the gun trigger. To make the spot weld, apply current and heat to the outer panel by releasing the electrode trigger. A spring will force the electrode against the outer panel, allowing amperage to heat the metal to a molten puddle. This should fuse the outer panel to the inner.

When the metal glows red to white hot, retract the electrode, but continue holding pressure on the gun and panel for 5 or 6 seconds until the hot spot cools. If everything was clean and fitted properly, you should have a good spot weld.

FRICTION STIR WELDING ALUMINUM

It is important to address this relatively new process because a lot is being said about the process in the automotive and aerospace news lately. The process is primarily being used to join aluminum sheets together that would usually be spot welded. The process is just as it is named, friction stirring the metal until it joins by heat generated by pressing and spinning a tool on two or more sheets of aluminum.

The tool to do this friction joining is even larger than the machine that is used to spot-weld with, and the tool must be programmed electronically to put the precise amount of pressure and spinning to generate the correct amount of heat in the sheets of aluminum. The process can also be used to join other metals, such as stainless steel and titanium,

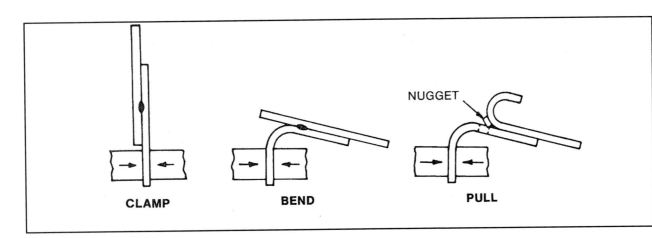

CLAMP   BEND   PULL

Test spot weld by clamping one piece in a vise and bending two pieces apart. If spot weld pulls apart, penetration is poor. If spot weld pulls a nugget—spot-welded metal—from one piece, penetration should be good. Courtesy The James F. Lincoln Arc Welding Foundation.

plus mild steel.

One of the notable projects to be certified for friction stir welding is the personal jet airplane made by Eclipse Aviation in Albuquerque, NM. The friction welding process is used to join the airplane skin to the forming bulkheads. But the process is much too complex to be considered for any small shop.

METAL SPRAYING

This process has been around for at least 75 years, but very little is commonly known about it. The first two uses of the metal-spraying processes were to apply hard-facing alloy steel layers to saw blades, mining equipment and to lathe tools, and the second use for metal spraying was to build up worn pulleys, oil seal surfaces and shafts. Almost any metal powder can be sprayed with the oxyacetylene heated torch.

More recent uses for metal spray are to fill cracks in cast-iron cylinder heads and cracks in cast-iron exhaust manifolds. The metal-spray equipment is easy to set up and to use. The photos above show two different setups and how they are used.

JEWELRY MAKING

Look at the chart on page 3 that gives the melting points of metals. You can see that gold melts at 1,946°F (1,063°C), and silver melts at 1,761°F (960°C). That is a higher melting point than the temperature that aluminum melts at—1,217°F (658°C)—so it shows that oxyacetylene will be a good heat source for welding rings, bracelets and other jewelry.

But don't use a full-sized gas welder to weld your wedding rings together! The amount of heat would quickly turn them both into a shapeless blob of gold! You need a tiny, little flame to accurately control the very small amount of heat necessary to do a super-delicate job like that. Conversely, the small jeweler's oxyacetylene torch could never weld the tubular framework on a race car or even a bicycle. Just remember the example of tack hammers and 10-lb. sledgehammers, related to railroad spikes and picture nails. It takes the right tool to do each job. Also, remember the lessons learned in chapter 1 about heat control.

Self-teaching yourself to do jewelry repair and jewelry making could be the start of a whole new career or a retirement income. But practice first before you try the wedding and engagement ring fusion.

COLD REPAIRS

Several years ago a motorcycle rider came into the welding shop with a pencil-diameter-sized hole punched in the transmission case of his expensive new motorcycle. He wanted me to weld it "at low temperature," he said. After I explained to him that his aluminum transmission case melted at 1,300°F, and that a welding temperature of at least that heat was required to fusion-weld the hole in the case, he asked what could be done to avoid taking the transmission apart for welding.

I suggested that he buy a package of epoxy-and-steel powder and repair the leak with J.B. Weld. He did as I suggested, and the "cold weld" is still holding. Try steel-based epoxy for similar metal repairs. Clear epoxy does not have the heat resistance or the flexibility that metal base epoxy does. You could save yourself a lot of work with the "cold weld" method. Try it sometime.

Chapter 16
WELDING PROJECTS

Welding supply dealers or steel supply dealers can sell you as much or as little angle, strap, U-channel or round stock as you need. If you're going to be serious about this, then you need to develop a good relationship with your local dealer. Courtesy Western Welding Inc., Goleta, CA.

This chapter should be the most rewarding chapter in the whole book. You spent all that money on welding equipment and your spouse wants to know what you plan to do with it. Now is the time when you can prove that the investment was worth it. The projects in this chapter are entirely within your skill level if you have studied and practiced, practiced, practiced the techniques in this book.

I start the project off with the welding and cutting table. It is the single most useful thing I have ever built for my shop, except for my 8-foot workbench. I have no doubt that all my go-karts, race cars and airplane projects would have been a lot easier and faster to build if I had built this table for my shop way back in my teenage years. So, build this table first, and it will help all your other projects along. As you can see in the construction and assembly pictures, you can build this table out on the driveway or on the shop floor. Then, when

it is finished, you can cut, fit, bend and weld all your future projects on the welding and cutting table.

In each of the projects, I list the welding and cutting processes recommended for fabricating the project. For instance, in Project #1, the welding and cutting table, I recommend the cutting torch and arc welding, although MIG or TIG welding could also be used. And remember, use good-quality welding rod. The projects deserve good-quality fabrication materials.

PROJECT #1
WELDING & CUTTING TABLE

It will probably take 40 to 50 hours for a beginner to build this welding and cutting table, but it is worth the effort.

Welding Process—Cutting torch MIG or arc welding.

Materials—Steel.

Cut the Parts to Fit—This is a more advanced project, so do your own trimming to make the lower frame and secondary parts. Cut the legs to length. Make the top frame of 1-1/4" x 1-1/4" angle. Make it square. Weld on the 3/8"-thick tabletop. Trim all other parts to fit. Leave the sides open or fill them in with 16-gauge (0.060") sheet metal. You can make doors and a catch tray for sparks from the cutting torch. Tack-weld the 1" x 1/4" straps in place so you can replace them easily when they get "ratty" from cutting-torch flame. Paint everything but the tabletop.
Note: Several pages in this book explain that you can make your welding table much larger if you need one that is; say 4' x 8'. Make your table as large as you need it to be.

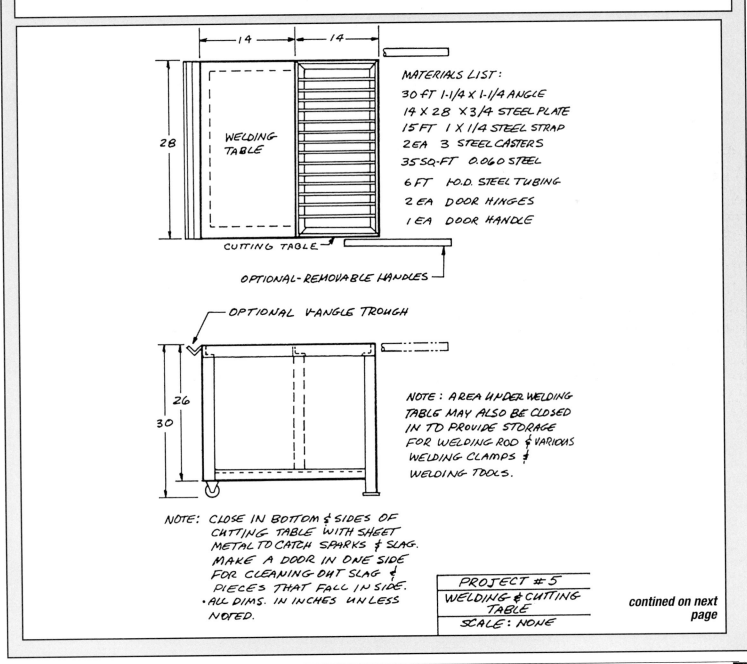

MATERIALS LIST:
30 FT 1-1/4 X 1-1/4 ANGLE
14 X 28 X 3/4 STEEL PLATE
15 FT 1 X 1/4 STEEL STRAP
2 EA 3 STEEL CASTERS
35 SQ-FT 0.060 STEEL
6 FT 10.D. STEEL TUBING
2 EA DOOR HINGES
1 EA DOOR HANDLE

WELDING TABLE

CUTTING TABLE

OPTIONAL-REMOVABLE HANDLES

OPTIONAL V-ANGLE TROUGH

NOTE: AREA UNDER WELDING TABLE MAY ALSO BE CLOSED IN TO PROVIDE STORAGE FOR WELDING ROD & VARIOUS WELDING CLAMPS & WELDING TOOLS.

NOTE: CLOSE IN BOTTOM & SIDES OF CUTTING TABLE WITH SHEET METAL TO CATCH SPARKS & SLAG. MAKE A DOOR IN ONE SIDE FOR CLEANING OUT SLAG & PIECES THAT FALL INSIDE.
• ALL DIMS. IN INCHES UNLESS NOTED.

PROJECT #5
WELDING & CUTTING TABLE
SCALE: NONE

continued on next page

The welding table project started with a trip to Western Welding, Inc., where they cut the metal to correct lengths for me, for about 1% of the total cost of material.

A fence stretcher come-along holds the table sides in place while I tack-weld them. This project is laid out on concrete.

Welding/cutting table legs, sides and top were arc-welded with a 225-amp buzz-box welder, using E-6010 arc-welding rod 1/8" diameter.

1/8" holes were drilled in framework and the aluminum sides were Cleco'ed in place.

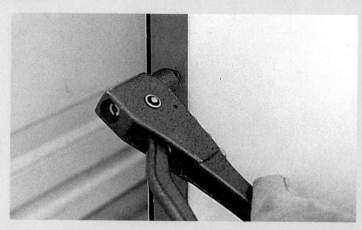

When all the 1/8" rivet holes were drilled and deburred, 1/8" steel pop rivets were installed to hold the aluminum sides on the table. Aluminum cannot be welded to steel, so I used rivets to fasten the aluminum sides, bottom and front to the steel framework.

This front quarter view of the completed table shows the class A B C dry chemical fire extinguisher mounted where it is easy to reach, and it shows the 4" bench vise mounted on the cutting table side.

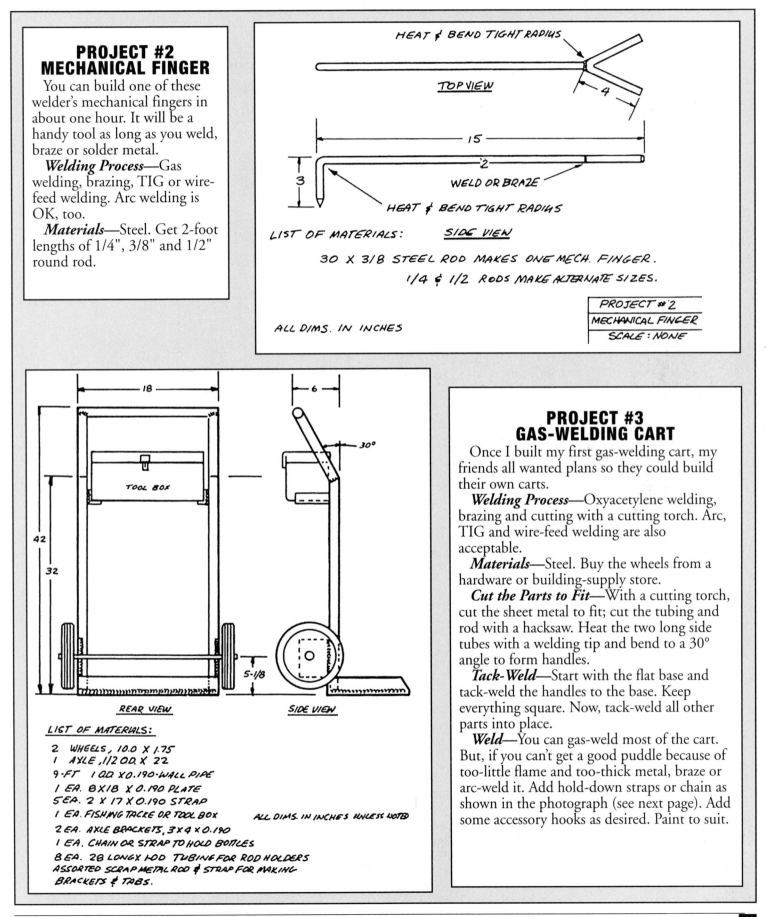

PROJECT #2
MECHANICAL FINGER

You can build one of these welder's mechanical fingers in about one hour. It will be a handy tool as long as you weld, braze or solder metal.

Welding Process—Gas welding, brazing, TIG or wire-feed welding. Arc welding is OK, too.

Materials—Steel. Get 2-foot lengths of 1/4", 3/8" and 1/2" round rod.

HEAT & BEND TIGHT RADIUS

TOP VIEW

4

15

2

WELD OR BRAZE

3

HEAT & BEND TIGHT RADIUS

SIDE VIEW

LIST OF MATERIALS:

30 X 3/8 STEEL ROD MAKES ONE MECH. FINGER.

1/4 & 1/2 RODS MAKE ALTERNATE SIZES.

ALL DIMS. IN INCHES

PROJECT #2
MECHANICAL FINGER
SCALE: NONE

18

6

30°

TOOL BOX

42

32

5-1/8

KENWOOD

REAR VIEW

SIDE VIEW

LIST OF MATERIALS:
2 WHEELS, 10.0 X 1.75
1 AXLE, 1/2 OD. X 22
9-FT 1 OD X 0.190-WALL PIPE
1 EA. 8 X 18 X 0.190 PLATE
5 EA. 2 X 17 X 0.190 STRAP
1 EA. FISHING TACKLE OR TOOL BOX
2 EA. AXLE BRACKETS, 3 X 4 X 0.190
1 EA. CHAIN OR STRAP TO HOLD BOTTLES
8 EA. 28 LONG X 1 OD TUBING FOR ROD HOLDERS
ASSORTED SCRAP METAL ROD & STRAP FOR MAKING
BRACKETS & TABS.

ALL DIMS. IN INCHES UNLESS NOTED

PROJECT #3
GAS-WELDING CART

Once I built my first gas-welding cart, my friends all wanted plans so they could build their own carts.

Welding Process—Oxyacetylene welding, brazing and cutting with a cutting torch. Arc, TIG and wire-feed welding are also acceptable.

Materials—Steel. Buy the wheels from a hardware or building-supply store.

Cut the Parts to Fit—With a cutting torch, cut the sheet metal to fit; cut the tubing and rod with a hacksaw. Heat the two long side tubes with a welding tip and bend to a 30° angle to form handles.

Tack-Weld—Start with the flat base and tack-weld the handles to the base. Keep everything square. Now, tack-weld all other parts into place.

Weld—You can gas-weld most of the cart. But, if you can't get a good puddle because of too-little flame and too-thick metal, braze or arc-weld it. Add hold-down straps or chain as shown in the photograph (see next page). Add some accessory hooks as desired. Paint to suit.

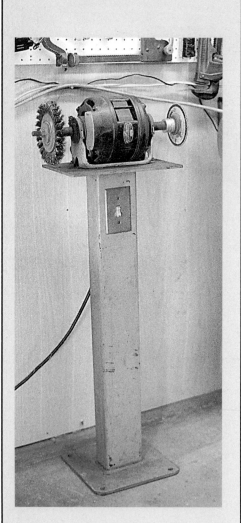

Completed gas-welding cart has been put to good use.

PROJECT #4
SANDER & GRINDER STAND

Build this grinder stand in one to two hours, then count the many hours of use it will see afterwards.

Welding Process—Weld with an arc welder and cut with a cutting torch.

Materials—Steel. You may also be able to pick up some of the materials at a scrap-metal yard.

Cut Parts to Fit—With a cutting torch, cut the pipe and plates to size.

Weld—Arc-weld the plates to the ends of the pipe. Paint to suit.

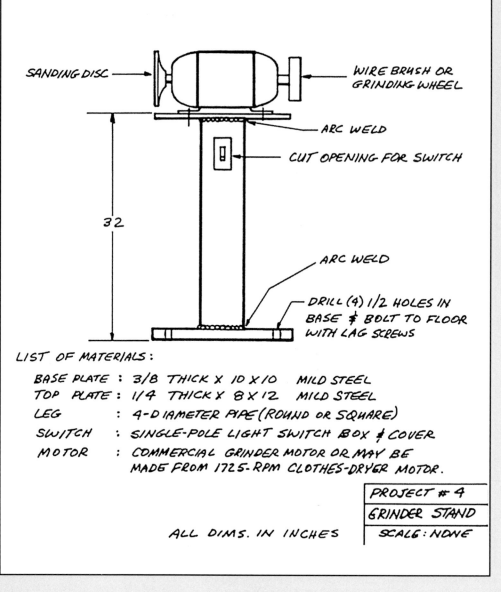

SANDING DISC

WIRE BRUSH OR GRINDING WHEEL

ARC WELD

CUT OPENING FOR SWITCH

32

ARC WELD

DRILL (4) 1/2 HOLES IN BASE & BOLT TO FLOOR WITH LAG SCREWS

LIST OF MATERIALS:

BASE PLATE : 3/8 THICK X 10 X 10 MILD STEEL
TOP PLATE : 1/4 THICK X 8 X 12 MILD STEEL
LEG   : 4-DIAMETER PIPE (ROUND OR SQUARE)
SWITCH   : SINGLE-POLE LIGHT SWITCH BOX & COVER
MOTOR   : COMMERCIAL GRINDER MOTOR OR MAY BE MADE FROM 1725-RPM CLOTHES-DRYER MOTOR.

PROJECT # 4
GRINDER STAND
SCALE: NONE

ALL DIMS. IN INCHES

PROJECT #5
JACK STANDS & WORKSTANDS

These shop-made workstands will come in handy when you start building the trailers shown in this chapter.

Welding Process—Arc welding, cutting torch.

Materials—1-1/2" to 2" pipe for center shaft. Metal plate is 0.190" to 0.375" thick, depending on which of the pictured stands you make.

Plan Before Cutting—When you decide which workstand to build, make a simple drawing like the ones in this chapter. Don't just start cutting and welding. Other people may want you to make workstands for them once they see yours.

Sandblast & Paint—This prevents corrosion and improves the project's appearance.

Note: All the projects listed in this chapter will benefit from pre-cleaning before welding. Sandblast or glass-bead clean the raw steel. Power-sand the joints to be welded before welding and then sandblast or wire-brush and power-sand the welds after welding. Make it real pretty!

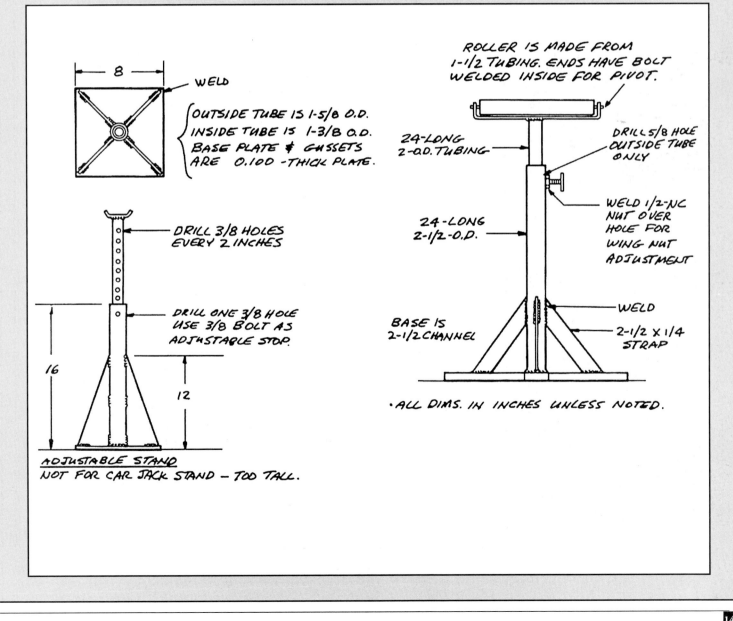

8

WELD

OUTSIDE TUBE IS 1-5/8 O.D.
INSIDE TUBE IS 1-3/8 O.D.
BASE PLATE & GUSSETS
ARE 0.100 -THICK PLATE.

DRILL 3/8 HOLES
EVERY 2 INCHES

DRILL ONE 3/8 HOLE
USE 3/8 BOLT AS
ADJUSTABLE STOP.

16

12

ADJUSTABLE STAND
NOT FOR CAR JACK STAND — TOO TALL.

ROLLER IS MADE FROM
1-1/2 TUBING. ENDS HAVE BOLT
WELDED INSIDE FOR PIVOT.

24-LONG
2-O.D. TUBING

DRILL 5/8 HOLE
OUTSIDE TUBE
ONLY

24-LONG
2-1/2-O.D.

WELD 1/2-NC
NUT OVER
HOLE FOR
WING NUT
ADJUSTMENT

WELD

BASE IS
2-1/2 CHANNEL

2-1/2 X 1/4
STRAP

·ALL DIMS. IN INCHES UNLESS NOTED.

PROJECT #6
TOW BAR FOR CARS & JEEPS

Welding Process—Cutting torch and arc, TIG or wire-feed welding.

Materials—Buy only good-quality new materials. Two tow-bar capacities are described. The light tow bar will safely tow cars up to 3,500 lbs total weight; heavy bar up to 6,000 lbs. My light tow bar has pulled my Corvairs across the U.S. and a Scirocco race car up and down the West Coast. If the tow bar is bolted snugly to the towed vehicle and careful towing habits are observed, cars will track well at all legal speeds.

Cutting and Fitting—A tow bar must be a perfect triangle to tow a car straight. Consequently, it's best to make a jig. Use a 1-7/8" hitch ball as the center point to line up the hitch and lay out a triangle on a piece of 1/2" plywood. Use a 1/2" steel bar as the base of the triangle. Tack-weld the tow bar in the jig, then remove it for final welding.

Attaching to Tow Vehicle—On my Scirocco race car, I removed the front bumper guards, exposing a bolt hole under each guard. In each hole, I put a 3/8" grade-5 bolt. I also drilled two more 3/8" holes on the bottom flange of the bumper for two more bolts—and the tow-bar angle. These four bolts and the two 1/2" hinge bolts were all double-nutted and checked every 100–200 miles of towing. The heavy-duty tow bar uses larger bolts, of course.

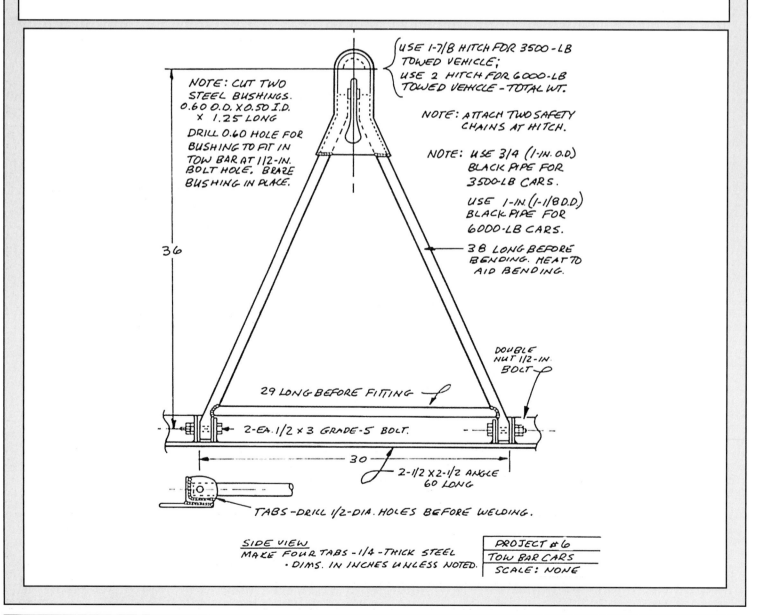

USE 1-7/8 HITCH FOR 3500-LB TOWED VEHICLE; USE 2 HITCH FOR 6000-LB TOWED VEHICLE - TOTAL WT.

NOTE: ATTACH TWO SAFETY CHAINS AT HITCH.

NOTE: USE 3/4 (1-IN. O.D.) BLACK PIPE FOR 3500-LB CARS. USE 1-IN. (1-1/8 O.D.) BLACK PIPE FOR 6000-LB CARS.

NOTE: CUT TWO STEEL BUSHINGS. 0.60 O.D. X 0.50 I.D. X 1.25 LONG DRILL 0.60 HOLE FOR BUSHING TO FIT IN TOW BAR AT 1/2-IN. BOLT HOLE. BRAZE BUSHING IN PLACE.

36

38 LONG BEFORE BENDING. HEAT TO AID BENDING.

DOUBLE NUT 1/2-IN. BOLT

29 LONG BEFORE FITTING

2-EA. 1/2 X 3 GRADE-5 BOLT.

30

2-1/2 X 2-1/2 ANGLE 60 LONG

TABS-DRILL 1/2-DIA. HOLES BEFORE WELDING.

SIDE VIEW
MAKE FOUR TABS - 1/4-THICK STEEL
• DIMS. IN INCHES UNLESS NOTED.

PROJECT #6
TOW BAR CARS
SCALE: NONE

PROJECT #7
TRAILERS FOR RACE CARS & AIRPLANES

Welding Process—Cutting torch, arc welding.

Materials—For the axle, I use the rear-axle assembly from a front-wheel-drive car. I just pick a wheel I like, and cut and splice or cut and narrow the axle to suit the desired dimensions. Springs can be purchased at a boat-trailer supply house or wheel manufacturer. Never build a trailer without springs. That $100 you might save by not buying springs could result in $1,000 worth of damage to whatever you're hauling due to the rough ride. For a light trailer, try the rear axle from a 1980-or-later GM X-Body (Chevy Citation, Buick Skylark, etc.). For a heavier-duty axle, use Cadillac Eldorado, Olds Toronado or 1979-and-later Buick Riviera rear axle.

Dimensions—These plans are for a trailer that will haul a 2,000-lb car. Scale up the dimensions for a slightly larger car or down for hauling smaller items such as a home-built airplane. The trailer tongue can be lengthened at least 10 feet for airplane towing.

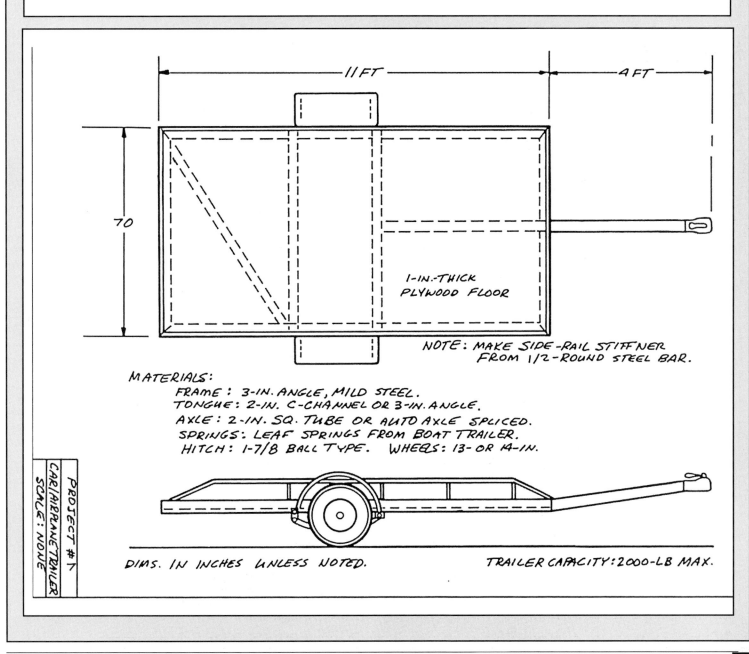

MATERIALS:
FRAME: 3-IN. ANGLE, MILD STEEL.
TONGUE: 2-IN. C-CHANNEL OR 3-IN. ANGLE.
AXLE: 2-IN. SQ. TUBE OR AUTO AXLE SPLICED.
SPRINGS: LEAF SPRINGS FROM BOAT TRAILER.
HITCH: 1-7/8 BALL TYPE.   WHEELS: 13- OR 14-IN.

NOTE: MAKE SIDE-RAIL STIFFNER FROM 1/2-ROUND STEEL BAR.

1-IN.-THICK PLYWOOD FLOOR

DIMS. IN INCHES UNLESS NOTED.   TRAILER CAPACITY: 2000-LB MAX.

PROJECT #7
CAR/AIRPLANE TRAILER
SCALE: NONE

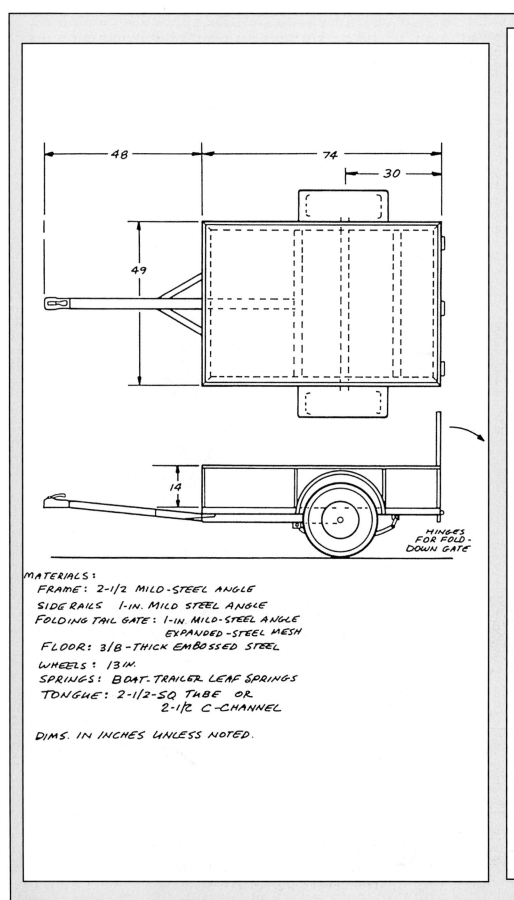

MATERIALS:
  FRAME: 2-1/2 MILD-STEEL ANGLE
  SIDE RAILS 1-IN. MILD STEEL ANGLE
  FOLDING TAIL GATE: 1-IN. MILD-STEEL ANGLE
                     EXPANDED-STEEL MESH
  FLOOR: 3/8-THICK EMBOSSED STEEL
  WHEELS: 13 IN.
  SPRINGS: BOAT-TRAILER LEAF SPRINGS
  TONGUE: 2-1/2-SQ TUBE OR
          2-1/2 C-CHANNEL

DIMS. IN INCHES UNLESS NOTED.

HINGES
FOR FOLD-
DOWN GATE

PROJECT #8 UTILITY TRAILER

These plans can be slightly modified to include a plywood floor or a full mesh floor and sides.

I purposely chose a Chevy Citation rear-axle assembly as the basis for this project because the Citation wheel bolt pattern is the same size as my Chevy Cavalier convertible that will tow the trailer. That way, my spare tire in the Cavalier convertible will fit the utility trailer if I ever have a flat tire on the trailer. Conversely, the trailer wheels and tires will also fit my car if I need an extra spare for it. It makes sense to standardize. Also, the trailer weighed 300 pounds when it was finished, very light and no problem for towing behind a compact car.

Welding Process—Cutting torch, arc welding on the frame, TIG welding on the axle. Plasma arc cutting for the smaller angle brackets and tail light holders.

Materials—The axle shaft was replaced with 2-1/2" x .125" wall square tubing. The tongue was made from 2" OD schedule 40 pipe. Other materials are given on the plans.

Variations—As shown in the photos, this trailer can be varied to carry 3 motorcycles for a steel bed lawn equipment or snowmobile trailer, or a plywood bed utility trailer. Be sure to add tail and stop lights. Get your trailers registered with your state Department of Motor Vehicles.

The trailer, Project #7, was laid out on a concrete driveway and welded with a Lincoln 225 stick welder and E-6011 and E-6010 welding rod.

Trailer frame and tongue is welded and ready for fitting the springs, axle and wheels.

To save weight and make a prettier trailer, I made an axle from 2" x 2" x .125" wall square tubing.

I TIG-welded the axle to wheel spindle adapters, using Cronatron #222T high-strength rod. Note the really pretty welds!

The frame is now painted white with black trim to match my tow cars, a Chrysler LeBaron and a Chevrolet Cavalier. The tail lights are wired and the bed is ready for the 5/8" outdoor plywood floor.

Late-night fabricating of wood fenders to be ready for Department of Motor Vehicles inspection and registration the next day. The trailer weighed 300 pounds at the DMV inspection. Cost with new tires was $300, or $1 per pound.

Chapter 17
WELDING CERTIFICATION AND TRAINING

This work truck is outfitted for welding and other kinds of maintenance and repairs. Somebody is making a living out of this "Welder's Office."

Now that you have digested the material in this book and practiced some of the techniques, it is now time to continue your education. There are many ways to make your living as a welder, if you choose to, or you can just pursue it as a hobby. For those wishing to further their skills and turn them into a career, there are several certification tests you can take, as well as welding schools you can attend.

Regardless of what you intend to do with welding, one of the best things you can do is seek out and study good welds. You can learn a lot about welding by looking closely at welds that are obviously very good, aircraft or race car quality. Look and take note of "pretty" welds, then practice, practice, practice until your welds look as good. That is one very good way to improve your welding skills. Try to be the best welder you can be.

CERTIFICATION

Certification, or qualification as it's sometimes called, is very helpful if you plan to make a living as a welder. However, you can weld professionally without being certified. Certification is something like having a college degree in sales. You can earn a living selling without it, but being certified may help you get a job by proving you are qualified. Regardless of certification, most large businesses and agencies conduct their own certification tests, or they will have an outside testing lab validate your certification tests.

Documentation

What do you have to show once you're certified? You'll be given a wallet-sized card listing the metals you are certified to weld and by what process(es). Also indicated will be whether you are a Class A (excellent), Class B (good) or Class C (OK) welder. Typically, the card and certification expire in 3, 6 or 12 months. You must be retested to stay certified.

You can be certified in different specialized areas: spot-welding contacts on home-heater thermostatic controls, TIG-welding stainless-steel pipe in the horizontal position in nuclear power plants, and so on. Two well-known certifications are for petroleum pipe-welding and aircraft TIG-welding. Certification testing is always conducted under strict supervision.

Pipe Welding Certification

The exam for this certification process is welding a section of 5/16" wall, 6"-OD pipe called a coupon. The ends of two 6"-long sections of pipe are beveled in a lathe and placed end-to-end for tack-welding. Usually, a backing ring is placed at the pipe joint with 1/8" spacers sticking out to obtain the proper root opening—distance between the pipe ends at the weld seam—to ensure that 100% penetration is obtained.

The pipe is placed at the welder's eye level. This positions the weld seam so all welding positions—horizontal, vertical and overhead—and combinations of each must be used to complete one pass. The welder starts with a root pass—the first bead. More than one pass is required to fill the weld seam. Except for the last one, following passes are appropriately called fill passes—usually one or two are required. The last pass is called the cap, or weld-out pass.

Slag must be removed after each pass to eliminate possible slag inclusions.

After welding is completed, a coupon from the weld seam is removed and scrutinized. Coupons are cut from the pipe with a saw or cutting torch. The weld bead is then ground flat and even with the pipe surface. Each coupon is placed in a test machine and bent backward to a horseshoe shape. Any porosity or cracking of the weld or base metal will fail the welder.

When certifying for arc welding, 5P welding rod is used for oil-field mild-steel pipe; E-7018 rod for 4130 steel pipe in steam and nuclear powerplant welding. TIG and wire-feed techniques are also used. As mentioned, the employer usually gives the certification test. However, there are also individual testing laboratories. A couple of testing laboratories are:

Advanced Testing Laboratories
4345 E. Imperial Hwy.
Lynwood, CA 92063

Atomic Testing Lab
5620 Modesto
Albuquerque, NM
Phone: (505) 828-2315

Aircraft Welding Certification

Several coupons—both tubing and sheet—must be welded successfully before you can become a Class A certified aircraft welder. Exact procedures may vary from country to country, depending on that government's regulatory laws. But basically, there are four groups of metals to be certified on:

• Group IA, 4130 Steel—One cluster, cross-sectioned, polished and then etched for examination.
• Group IIA, Stainless Steel—Butt-weld 0.032" and 0.063" plates together. Complete fusion is required. Weld is sectioned, ground and bent. Visual inspection follows.
• Group IV, Aluminum—Butt-weld 0.032" and 0.063" plates together. Complete fusion is required. Weld is sectioned, ground and bent. Visual inspection follows.
• Group VI, Titanium—Special request; test production part by sectioning, polishing and 10X magnification inspection.

In the previous two versions of this book, there was a section that covered welding 3 tubes of 1" diameter x .063" wall to a plate that was .125" thick to pass certification for aircraft welding. That test has been changed to a more useful and practical test. The new test simply requires the welder to weld up a small cluster of 4130 steel tubes that represent the kind of welds that he or she usually does on a regular basis when doing repair welds or when manufacturing new tubular structures for a manufacturer of airplane parts.

A typical welder might be doing certified repairs on cracked engine mounts and would need to be certified to weld this kind of tubing. Therefore the welder would make a tubular coupon that would be made from 1" diameter, .049" wall tube with a 7/8" diameter tube x .035" wall, fitted and welded to the 1" tube at a 90-degree angle and another tube of 5/8" diameter x .035" wall at a 45-degree angle fitted to the other two tubes and welded together.

For the test requirements for aircraft and aerospace welding, contact the following and request the following specifications:

The American Welding Society
550 NW LeJeuene Rd
Miami, FL 33126-5671
Phone: (305) 443-9353

Society of Automotive Engineers (SAE)
400 Commonwealth Drive
Warrendale, PA 15096-0001
Phone: (724) 776-4841

Request the following specifications :
DIN 29591 (German)
AS7110/5B Aerospace Standard
AS7110B Aerospace Standard
AWS D17.1:2001 American National Standard

Specification for Fusion Welding for Aerospace Applications

DIN 29591: Performing the welder qualification tests
AS7110/5: Accreditation Program, requirements for fusion welding
AS7110, rev B Accreditation Program for Brazing

Aluminum Butt Plates Test Procedure—

Minutes prior to actually making the TIG weld, a Scotchbrite abrasive pad should be used to slightly roughen each piece of aluminum about 1/2" and a 1/4" air space should be left under the weld seam. This ensures that 100% penetration is achieved. If the bead is not thicker than the base metal on both sides, you fail the test. Even though I don't use back gas, I use my stainless-steel back-gas clamping fixture to clamp the aluminum.

When the testing lab checks the aluminum weld coupon, they sand the weld bead to the thickness of the base metal, cut the coupon into strips and pull-test them. The aluminum strips must not break in the heat-affected area. Also, craters and porosity are not allowed.

Stainless Steel Butt Plates Test Procedure— Stainless steel coupons are always TIG welded with back-gas purging to prevent crystallization on the back side of the bead. Clamping and back-gas purging are done with the fixture pictured above.

Prior to welding, clean the stainless-steel plates with MEK or alcohol. Use clean, high-quality welding rod. To ensure that argon gas covers the weld while solidifying, keep the torch over the bead after completion until the purge-gas stops. The timer on the TIG machine usually provides 5-6 seconds of gas flow after the torch is off.

The testing laboratory sands the stainless steel weld down to base-metal thickness, then cuts it into test strips. The strips are then bend-tested at the weld seam. Without 100% penetration, the bend test will break through at the bead. If so, the weld will fail.

Welding Titanium—This is an on-the-job test that uses an actual production part. The part is then destruction-tested. It is sectioned through, pulled and bent, and subsequently polished and inspected with a 10X magnifying glass.

WELDING FOR A LIVING

I once saw a bumper sticker that read, "Welding holds the world together." Welding actually does hold more things together than most people realize. For instance, your automobile has hundreds of spot welds, seam welds and weld beads. Take a drive and you'll see steel bridges with thousands of welds. You'll pass by skyscrapers that are welded-steel structures underneath all that brick, mortar and glass. People welded those things together. Even robot welders have to be set up and adjusted by a welder—er, uh, welding technician—so the welds are as good as those welded by human hands.

Ordinarily, welding pays as well as most skilled trades. In many cases, a welder's salary matches that of a degreed engineer. Likewise, a welding engineer is one of the highest-paid engineers. The highest-paying jobs for welding engineers are in offshore structures and structural-steel industries. Why? The forces of supply and demand govern here. Welding is a skilled trade and there is an element of danger involved.

Welding is like riding a bicycle; once you learn how, you never forget. But you do need to practice from time to time. Once your friends and neighbors discover that you can weld, you'll be asked to fix many things—now that will keep you in practice. You could even make a little extra money.

WELDING SCHOOLS

Some colleges and industrial arts schools provide evening and weekend classes for welders desiring to become certified. Trade schools also provide classes for welding certification. In these classes, a student is first taught the basics on pieces of scrap metal, and then given several weeks of nothing but practice, practice, practice to improve skills.

If you are serious about wanting to become a professional welder, start by taking a welding class. Most high schools teach welding as an elective. And most two-year colleges provide a degree in welding technology. It's interesting to note, though, that many four-year colleges and universities do not provide welding courses or degrees. Those that do usually are engineering schools.

Most welding manufacturers, such as Lincoln Electric Co. and Hobart Brothers, operate welding schools at their factories. Lincoln's school, the James P. Lincoln Arc Welding Foundation, provides scholarship awards for students of other colleges and universities. For more information contact:

Aluminum Company of America (ALCOA)
Technology Marketing Division
303-C ALCOA Building
Pittsburgh, PA 15219

Hobart Brothers
Box HW-34
Troy, OH 45373

The James P. Lincoln Arc Welding Foundation
Box 17035
Cleveland, OH 44147

"The average welder in the U.S. is 54 years old...there could be a shortage of 200,000 skilled welders by the year 2010." — *Welding Journal*, American Welding Society. January 2007 issue.

GLOSSARY

Air-Arc Gouging—An electric-arc process that cuts metal by melting it with a carbon or copper electrode, and simultaneously blows away the molten metal with a 100 psi air blast through the center of the electrode. It's a very noisy and messy process, but gets the job done cheaply where a lot of metal has to be removed.

Alloy—Basic metal modified by chemical compositions to improve its hardness or corrosion-resistant characteristics.
Aluminum Heat-Treating—Process by which aluminum is heated to 960°–980°F (516°–527°C), then quickly cooled, or quenched. The temperature is reached by placing the part in an oven or in a bath of liquid salts called solution heat treatment. Quenching is accomplished by using cool air or water. Quick cooling is the secret to heat-treating.

Annealing—Opposite of hardening. It's done to remove hardness in certain metals where drilling or other machining is desired. The metal is usually heated to about the same temperature as for heat treating, but then allowed to cool slowly. Aluminum and steel may be annealed. Usually, the part can be heat-treated again after annealing.

Arc Blow—Deflection of the arc from its normal path by magnetic forces, usually associated with d-c welding.

Arc Welding—A welding process that fuses metal by heating it with an electric arc and simultaneously depositing the electrode in the molten puddle.

Backfire—Momentary, loud pop at the oxyacetylene torch tip. It is caused by the flame backing up into, or combustion occurring inside, the tip. It's usually a result of the weldor trying to get more heat from a torch with low gas pressure by holding it too close to the work and overheating the tip. This occurs more readily in a large torch with a rosebud tip because of low gas pressure. Backfire can be dangerous and should not be allowed to continue.

Backhand Welding—Like walking backward, it is welding backward. The weldor points the torch at the already welded seam, away from the unwelded seam. I doubt the need for this procedure, but it could be useful to avoid burning through very thin metal. The added mass of weld bead could help absorb the extra heat.

Backing Ring—Metal ring placed inside the seam of pipe being butt welded—welded end to end. The ring provides for full weld penetration and 100% strength in butt welded pipe seams. It is usually tapered for smooth flow of liquids, steam or gas once the weld is completed. The ring is then left inside the pipe

Backing Strip—Metal strip that serves the same purpose as a backing ring—to ensure 100% strength and weld penetration.

Bare Electrode—A consumable, bare electrode used in arc welding with no flux coating.

Base Metal—Prime metal to be welded, brazed or cut. In auto bodywork, the car's steel fender is the base metal and the welding rod the filler metal.

Bead or Weld Bead—Result of fusing together a seam in two or more pieces of metal with welding rod. Usually, the bead is thicker than the base metal.

Bevel—Preparatory step prior to welding, whereby the edges of the base metal are angle-filed or ground to better accept the filler metal. When welding thick metal the bevel forms a V-groove that promotes better weld penetration.

Billet—A solid bar of metal usually made by forging rather than casting.

Braze—Non-fusion weld produced by heating a base metal above 800°F (427°C) and using a non ferrous filler metal. The liquid (above 800°F) filler metal flows between closely fitted surfaces of the metal joint by capillary action. The base metal is not melted in braze welding.

Butt Joint—Joint between two pieces of metal lying flat, end to end.

Capillary Action—Action whereby the surface of a liquid—including metal heated to a liquid state—is raised, lowered or otherwise attracted to fixed molecules nearby. A plumber sweats together copper pipe by using liquid brass and other liquid solder to flow into tight-fitting places by capillary action. Dictionary defines it as "the force of adhesion between a solid and a liquid."

Carburizing—1. A heat-treating process that hardens iron-based alloys by diffusing carbon into the metal. The metal is heated for several hours while in contact with carbon, then quenched; 2. In gas welding, an acetylene-rich flame that coats the metal with black soot.

Corrosion—Gradual chemical attack on metal by moisture, the atmosphere or other agents. This includes rust on iron or steel, oxidation on aluminum, and acid pitting and etching on stainless steel. Corrosion is the biggest long-term problem for fabricated metal construction. Corrosion prevention can be accomplished by painting, plating, oiling or any other coating that keeps oxygen away from the base metal.

Cover Glass—Clear glass used in goggles and welding helmets to protect the more expensive colored lens from weld spatter.

Covered Electrode—Arc-welding electrode, used as a filler metal, that is covered with flux to protect the molten weld puddle from the atmosphere until the puddle solidifies. Commonly called slick electrode.

Deposit—Filler metal added during the welding operation.

Depth of Fusion—Depth that fused filler material extends into the base metal.

Duty Cycle—Seldom-understood term that applies to electric-arc welders, not gas welders. The duty cycle is a ten-minute period. If an arc welder has a 100% duty cycle, it can be used 10 minutes out of every 10 minutes or 100% of the time. If a welder has a 20% duty cycle, it can be used two minutes and must cool off for eight minutes out of each 10 minutes. Usually, a small arc welder will have a 90% duty cycle at the lower amp settings, tapering off to 5% at the highest settings.

Dye-Penetrant Testing—An inexpensive process to check welds for cracks and other defects. The process consists of three chemicals in solution: a cleaner, a spray-on or brush-on penetrating red dye, and a white developer solution. After the area is cleaned and the red dye allowed to soak in a few minutes, the developer is sprayed on. Defects show red and smooth areas appear white. After inspection, the cleaning solution can be used to remove the dye.

Field Weld—Weld done at the site or in the field rather than in a welding shop.

Filler Metal—Welding rod or other metal added to the seam to assure a maximum thickness weld bead.

Fillet Weld—Weld deposit of filler metal approximately triangular in shape. Usually made when welding a T-joint or 90° intersection.

Flash Burn—Burn caused by ultraviolet-light radiation from the arc in arc welding. Usually painful and more severe than sunburn, especially when the eyes are flash burned.

Flashback—Burning of mixed gases inside the torch body or hoses. Usually accompanied by a loud hiss or squeal. Must not be allowed to continue! Shut off immediately if flashback occurs, then shut off acetylene. Use in-line arrestors to prevent flashback. If flashback occurs, do not light up the torch again until you find the cause and eliminate it.

Flux—Chemical powder or paste that cleans the base metal and protects it from atmospheric contamination during soldering or brazing. Flux consists of chemicals and minerals that properly clean and protect each type of metal. Therefore, each type of metal joining requires a specific formulation of flux. Flux is not used in fusion welding, except as a coating over arc-welding rod or in submerged arc welding.

Fusion Welding—The only true kind of welding. The metal pieces to be welded are heated to a liquid state along the weld seam, and usually filler metal of the same or similar type is added to the molten puddle and allowed to cool, forming one continuous piece of metal. Thus, the weld should be stronger than the added filler metal.

Gas Metal-Arc Welding—Or *GMAW*, a process in which an inert gas such as argon, helium or carbon dioxide is fed into the weld to shield the molten filler metal. This displaces atmospheric air and inhibits oxygen from combining with the molten metal and forming oxides and other impurities that would weaken the weld. Also known as metal inert gas—MIG—welding, but commonly called wire-feed welding.

Gas Tungsten-Arc Welding—Or *GTAW*, another type of inert-gas arc welding with tungsten as the electrode material. Tungsten is used because it will not melt at welding temperatures. The arc is similar to the heat from an oxyacetylene torch except that it can be concentrated in a much smaller space. Helium or argon gas is used to shield the weld puddle—argon is preferred. The filler metal is uncoated. This process can be used to weld steel, stainless steel titanium, aluminum, magnesium and several other metals. Also known as tungsten inert gas—TIG—welding and commonly known as Heliarc welding. Heliarc is a registered trademark of Linde Corp.

Gas Welding—Also known as oxyacetylene welding. Common term to describe welding accomplished by burning oxygen and acetylene to make a 6,300°F (3,482°C) flame.

Hard Facing—Applying a very hard metal face to a softer metal to improve wear characteristics, such as on a bulldozer blade. The welding rod itself is the hard metal.

Hardness Testing—There are three types of tests: Rockwell, Brinell and Shore. Rockwell Hardness is a two-stage ball-impression test. Brinell Hardness is a one-stage ball-indentation test. Shore Hardness is a drop test for indenting metal. All three hardness tests use the principle that the harder the metal is, the less likely it is to be dented by a given force. Obviously, all three methods require calibrated special equipment, and a trained operator to analyze the results.

Heat-Affected Zone—Portion of the base metal that has not melted, but that has become discolored or blued by the heat from welding or cutting. Usually, metal strength is changed in the heat affected zone. The heat-affected zone is usually detectable by eye.

Heat Sink—A mass of metal, water-soaked rag or other heat absorbing material, placed so it absorbs heat, thus preventing overheating of a component or area. The use of a heat sink in welding can prevent or limit burn-through or warpage.

Heat-Treating—A process that adds strength and brittleness to metal. Almost all metals have a critical temperature at which their grain structure changes. This involves controlled heating and cooling of the metal to achieve the desired change in crystalline structure. Not all metals can be heat-treated.

Heliarc—Trademark of Linde. See TIG and Gas Tungsten Arc Welding.

Horizontal Position—Weld seam is horizontal.

Interpass Temperature—When several passes or beads are made in welding a joint, the lowest temperature of the weld bead before the next pass is started. Keep it as low as possible when welding cast iron to prevent cracking of the weld.

Joint—Junction where two or more metal pieces are joined by welding, brazing or soldering.

Kerf—Width of the cut, in oxyacetylene or plasma-arc cutting.

Keyholing—Usually occurs when butt-welding two very thin pieces of aluminum. A small hole melts all the way through but is filled with filler rod.

Magnaflux—A magic word in the welding or metallurgy business. An inspection process used with magnetic (ferrous) materials to detect cracks or other flaws. A fine iron powder is sprayed over the inspection area, then a strong magnetic field is induced electrically to cause any crack or defect to show as a separation of the iron particles. In most cases, this process causes the part to become magnetized, requiring demagnetization after the inspection is complete. Industrial Magnaflux equipment is similar in size and cost to a large arc welder.

Melting Point—Temperature at which a metal melts and becomes liquid. The melting point of frozen water (ice) is 32°F (0°C). Mild steel's melting point is 2,700°F (1,482°C).

MIG Welding—Metal inert-gas, or wire-feed, welding. The "M" for metal is a spool of wire fed through a shield of inert gas, usually 25% helium and 75% argon. It produces a result similar to tungsten inert-gas—TIG—welding. In the case of MIG welding, the filler rod or wire becomes the electrode and melts. Therefore, it must be continuously fed to the weld puddle. Stainless steel, steel, aluminum and other metals that can be fusion-welded can also be MIG-welded.

Neutral Flame—An oxyacetylene flame with equal amounts of oxygen and acetylene.

Nitriding—A surface-hardening process for certain steels, whereby nitrogen is introduced in contact with anhydrous ammonia gas in the 935–1,000°F (502°–538°C) range. Quenching is not required. It's accomplished in similar manner to carburizing.

Normalizing—This process is usually used on high-strength steels such as 4130 to remove strains in fabricated parts or in material intended for bending or machining. The metal is heated to a point above the critical transformation temperature and then allowed to cool in still air at room temperature—with no drafts.

Overhead Position—Welding position in which the seam is welded from the underside of the joint.

Oxidize—1. Effect of applying excess oxygen, causing metal to vaporize during welding, as in an oxidizing flame; 2. Slow chemical process whereby oxygen and water combine to attack ferrous metals, resulting in rust and corrosion.

Oxidizing Flame—Gas-welding flame with excess oxygen.

Oxygen Cutting—Special cutting process that cuts metals by the chemical reaction of oxygen and the base metal at elevated temperatures. No acetylene is used except to start the oxidizing process, then only oxygen is used.

Peening—Working metal with a small pointed hammer or spraying with small steel shot. Peening is usually done to improve the surface strength of the metal by putting it in compression. This prevents cracks from starting at the surface.

Penetration—Depth of weld metal from molten puddle into the base metal. Ideal penetration is 15% to 110%. Less penetration makes a weak weld. At more than 100%, the weld bead is thicker than the base metal.

Pickling—Just like putting cucumbers in solution, you put metal in a diluted acid or other chemical to clean oil, scale and other unwanted matter from its surface. Usually, a corrosion inhibitor, such as wax, is applied to the surface after pickling to prevent corrosion until the metal is painted or plated.

Plating—Outer coating of chromium, copper, nickel, zinc, cadmium or other heavy metal to enhance appearance or inhibit corrosion of parent metal. Usually, plating is accomplished by immersion in an acid solution with cathode and anode electric current, causing the

plating material to deposit on the parent metal. Most ferrous metals can be plated.

Plug Weld—Also called *rosette weld.* A circular weld made through a hole in one piece of tubing or blind channel, connecting another piece slipped inside.

Polarity—In direct-current arc welding, TIG-welding and wire feed welding, how current flows either positive to negative, or negative to positive. AC welding has no polarity because it switches between positive and negative polarity and back again 60 times per second. Heat is about equal at the electrode and workpiece. DC straight polarity (DCSP) is also called negative polarity. In DCSP, the workpiece is positive and the electrode negative. Heat is greater at the workpiece than at the electrode. DC, reverse polarity (DCRP) is also called positive polarity. In DCRP, the workpiece is negative and the electrode positive. Heat is greater at the electrode than at the workpiece and penetration is shallow.

Postheating—Heating a weld after it is completed, usually for stress relief.

Preheating—Heating the weld area beforehand to avoid thermal shock and thermal stresses. Metal is more sensitive to temperature than you would think. For instance, a steel pipe welded at 32°F (0°C) will be more brittle than one heated to 90°F (32°C), or even to 300°F (149°C). The larger the mass to be welded, the more it needs preheating.

Puddle—Liquid area of the weld where heat is being applied either by flame or electric arc. This is the most important part of welding! If the puddle is properly controlled, the weld will automatically be good.

Reducing Flame—In oxyacetylene welding, a flame with excess acetylene, imparting excess carbon to the weld. Also called a carburizing flame.

Residual Stress—Stress remaining in a structure after the weld joint cools. Metal wants to warp when it is heated. If a very strong jig or a triangulated

structure prevents warping, stresses remain in the area near the weld.

Root Opening—Distance between two pieces of metal to be joined.

Root of Weld—Point of weld farthest from heat source. Intersection point between bottom of weld and base-metal surface.

Rosette Weld—See *Plug Weld*

Sandblasting—Fast, easy method of cleaning certain metals before welding and painting. A high velocity air blast, carrying sand, is directed at the metal and the particles of sand abrade its surface. Obviously, this cleaning process should be used with caution to protect eyes and lungs. It leaves a rough surface and cleanup is messy.

Seam Welding—A form of spot welding in which two pieces of sheet metal are resistance-welded in a continuous seam.

Shielded-Metal Arc Welding—SMAW, also commonly known as stick welding or as the layman knows it—simply arc welding. Shielding is the flux coating on the metal rod.

Shop Weld—To prefabricate or weld subassemblies in a shop or controlled environment before taking them on-site for final assembly. Often used in large welding projects such as oil-drilling sites and nuclear power plants.

Slag—Impurities resulting from heating metal and boiling off dirt and scale present in most open-air welding. Slag is found at the kerf from oxyacetylene torch cutting. Slag will also be found in the hardened flux on top of an arc-welded bead.

Soldering—Metal joining process similar to brazing. Metal pieces are joined with molten solder, without melting the base metal. Solder is drawn into the joint by capillary action. As it cools, it sticks to the base metal. If two pieces of lead solder were joined by melting them together, technically that could be called welding. But I would just call it melting lead!

Spatter—Small, unsightly droplets of metal that deposit alongside the weld bead. Especially common in arc welding with E-6011 rod.

Spot Welding—A production welding method to join sheet metal. Electrical resistance heating and clamping pressure are used to fuse panels together with a series of small "spots." Filler metal is not used. Also called *resistance welding*.

Steel Heat-Treating—Process of heating and rapidly cooling steel in the solid state to obtain certain desired properties: workability, microstructure, corrosion resistance, and so on. Depending on its mass and alloy type, the steel is oven-heated to 1,475–1,650°F (802°–900°C), then quenched by dipping it in water or oil.

Stickout—Length of electrode (tungsten or wire) that sticks out past the gas lens, cup or gun.

Stitch Weld—Tack-welding technique with short weld beads about 3/4" long, spaced by equally long gaps with no welding. Used where a solid weld bead would be too costly and time-consuming, and where maximum strength is not required.

Stress Cracking—Metal cracking at the weld due to temperature changes or molecular changes. Overheated welds are more prone to stress cracking than underheated welds.

Stress-Relief Heat-Treating—When a complicated, rigidly braced structure such as an airplane engine mount or race-car suspension member is welded, stresses remaining in the metal will cause premature fatigue cracking unless they are relieved. Stress relieving is accomplished by heating part or all of the structure to about two-thirds of the melting point and then cooling it slowly. This allows the molecules in the structure to relax and stay relaxed.

Stringer Bead—A straight weld bead made without oscillation.

Submerged-Arc Welding—SAW or sub-arc welding. A process in which the electric arc is submerged in powder flux, thereby protecting the weld from atmospheric contamination. The system is usually automatic feed and travel and a base-metal rod is used. It's used where high accuracy and weld quality are desired.

Sugar—Crystallization in a weld. It usually occurs when welding stainless steel if the back side of the weld seam is not protected by an inert gas such as argon. Sugar has no strength and should not be allowed in a weld.

Tempering—When metal has been hardened by heat-treating, it usually becomes brittle. In order to relieve the internal strains, the metal is usually reheated to about one-quarter or one-half the temperature originally used in heat-treating.

Thoriated Tungsten—Tungsten electrode with 1–2% thorium added to provide a more stable arc. Thoriated tungsten is used to weld steel. However, pure tungsten must be used to weld aluminum and magnesium.

TIG—Tungsten inert-gas welding. Often called Heliarc, because helium was first used as an inert gas for this welding process. Argon and other gas mixtures are now used. Electrode is tungsten because it doesn't melt at welding temperatures. The inert gas shields the weld from atmospheric impurities, providing a high-quality weld. Filler rod is hand fed to the weld.

Tungsten Electrode—A non-consumable electrode used in TIG welding. The melting point of tungsten is 5432° F (3000C).

Ultrasonic Testing—Process to test metal parts for defects. High frequency sound waves are directed at the part, and their reflections are picked up by a receiver. Cracks and flaws inside the metal are detected by discontinuities in the return sound. Ultrasonic test equipment is expensive and requires trained personnel to operate and analyze the results.

Vertical Position—Type of weld in which the metal to be welded is vertical and the weld bead progresses upward or downward.

Weave Bead—Weld bead made with a transverse oscillation such as a figure-8, or a "Z" motion while moving forward along the seam. This bead deposits more filler metal and ties the two pieces together more effectively than a straight bead, but also provides the possibility of including slag or flux in the weld, thereby contaminating it.

Weld—Local melting together and fusing of metal produced by heating the base metal and, in most cases, applying filler rod to the molten puddle. The filler rod usually has a melting point approximately the same as the base metal, but above 800°F (427°C).

Weldability—Capacity of specific metals to be welded and to perform satisfactorily for the intended service. Not all metals are weldable. See Chapter 1 for what can and cannot be welded.

Weld Puddle—In welding, metal that is at or above its melting point and in a liquid state.

X-Ray—Inspection for stress cracks, internal corrosion or other defects. An actual X-ray picture is taken of the part. The operator must be specially trained because of the radioactive materials involved. This process is used to inspect welded seams on high pressure nuclear powerplant piping, airplane parts and other situations in which any defect would cause an expensive or potentially dangerous problem.

TERMS FOR WELDING DEFECTS

Arc Strike—Unintentional arc start outside of the weld bead. Usually more of a problem in TIG-welding in which strict quality control standards are observed.

Cold Weld—Poor penetration of the weld bead, usually less than 5% of the bead thickness.

Crater—In arc-welding and TIG-welding, a depression at the end of the weld bead caused by stopping the weld with too much heat applied.

Crater Crack—Crack in the crater at the end of the TIG-weld bead caused by stopping the weld with too much heat applied and with drawing the shielding gas before the weld solidifies.

Discontinuity—An interruption in the basic weld bead, usually excess filler material, but not necessarily a defect.

Drop Through—Filler material that sags through on the underside of the weld, caused by either too much heat or poor joint fit.

Inadequate Penetration—Depth of filler metal is less than 15% of the weld-bead thickness.

Porosity—Usually, gas pockets caused by the wrong weld temperature or a dirty, contaminated weld. Most porosity is caused by getting the weld bead too hot.

Slag Inclusion—Dirty weld due to flux trapped in the weld bead or scale or dirt from the base metal or welding rod.

Undercut—Cutting away of the base metal by improper application of temperature. The weldor just pointed the heat at the weld bead and capillary action pulled the molten puddle away from the cooler base metal to the hotter molten puddle. Undercut can be avoided by paying more attention to temperature control.